THEY

EXPLOSIVE UNDENIABLE SCIENTIFIC FACTS

CONNED

PROVING THE BIG BANG AND EVOLUTION IMPOSSIBLE

YOU

COLIN B NOBLE

Copyright © 2020 Colin B Noble

All rights reserved.

ISBN: 978-0-6488362-1-6

DEDICATION

To the youth of the world who have been indoctrinated into believing the theory of evolution is true science by mainstream education institutions and educators who have taught flawed illogical theories, misleading claims and endless unproven assumptions in the promotion of evolution and their religion of atheism.

CONTENTS

	Introduction.	1
1.	Your Journey of Discovery Begins.	7
2.	Where did the Theory of Evolution come from?	29
3.	The Big Bang.	75
4.	How reliable are dating methods for the Big Bang and Evolution?	105
5.	Undeniable Scientific Evidence Evolution is a con.	127
6.	Science of DNA proves Evolution possible.	153
7.	Atheist Religion and Evolutions War against Creationists.	171
	Epilogue.	201

INTRODUCTION

If you believe in The Big Bang and Evolution theories, you have been conned.

Yes, I admit this is a strong statement, especially when so many scientists claim their theories are true and their version of science conclusively proves it actually happened.

And that is the heart of the issue, has science categorically proven the Big Bang happened and we are all here today through the process of evolution.

The answer is a **BIG DEFINITIVE NO**, that's why they are called theories. The Big Bang and evolution are NOT proven science and those in the scientific community promoting evolution use untrue and misleading claims, and highly flawed processes designed to intentionally con the public into thinking they are similar to the 'law of gravity', something that is beyond question.

This book exposes their tactics, agenda, and motivations behind their Big Bang-Evolution claims and irrefutably proves it's nothing more than a big con fostered on the masses.

You will discover how the Big Bang and Evolution are not only impossible, they are built on highly flawed and unproven theories that are constantly being changed and modified as new proven scientific facts emerge that continually expose evolution's flaws and impossible claims.

I will present proven Biological and scientific facts that completely destroy the theory of evolution and expose the truth behind the evolution movement's agenda.

My journey in writing this book developed from researching evolutionists' rash claims and alleged supporting science. The more I learned, the more I unearthed a litany of false highly theoretical assumptions irresponsibly presented as truth. While researching the basis of their claims I discovered their many theories to be unsupported, illogical, with some being just plain crazy to the point where they needed to be exposed.

As an example, have you ever questioned the dating figures we frequently hear mentioned in nature programs or when visiting natural wonders? I even saw a real estate company promoting a farm for sale that they claimed contained rock formations 250 million years old. Where do they get that number from? Miraculously, there appears to be a date for every rock, geographical area, and even species, throwing around dates of 200 million, 350 million, 2 billion, 14 billion. Are these dates accurate? How do they arrive at these dates? Is there factual evidence to validate the claims, and what is the supporting science proving these dates are true and reliable?

My research uncovered a shocking truth behind dating methodology, and what you will discover will astound you. Their dating methods are unreliable, thus making their alleged supporting science weak and built on highly flawed assumptions. Scientists know this, yet they still throw out figures dogmatically as if they are proven scientific facts in a desperate attempt to support their flawed evolutionary theories. We will also expose the complicated and costly process involved in dating rocks and artefacts, shedding doubt on, if in fact, the claims made really have been dated.

Their unproven unsupported theoretical claims on all the wonders of this earth, it's creatures and the complexity of the human body, all evolved from a big random explosion, leading to a blob of stuff randomly forming and undergoing untold billions of purely random uncontrolled changes that had to occur in a perfect order for it all to happen, is impossible and cannot possibly be true. In reality, it's merely science fiction as you will soon discover.

My journey into the scientist's and theoretical physicist's world of the Big Bang and evolution theories was extremely enlightening. I learned that what these scientists claim as true is not truth. Their endless and differing theories are built on either unproven assumptions, or science so flawed one soon uncovers a path of deception orchestrated by the evolution scientists.

What evolution scientists and education institutions teach are merely highly flawed academic theories, resulting in one of the biggest cons fostered on mankind by the scientific community.

Yes, these are harsh and explosive words which will send evolution scientists into a tizzy, causing them to object and strenuously claiming their claims are true, it's science.

What they don't do is explain their definition of truth and science. It's certainly not the same as what we understand as truth. You will discover an evolution scientist's definition of truth is distorted and removed from reality.

My objective was to unquestionably prove the case 'Evolution is a Con', supported by clearly presented facts and reasoning in a way that a normal person like myself can relate to, free from confusing academic misleading jargon and theoretical arguments.

I'm not an academic, I'm an analytical researcher with a gifting of being able to clearly explain complex subjects in a common-sense practical way anyone can relate to. This book is designed to make a heavy, and at times complex, topic enjoyable, easy reading, and suitable for students to counter the misinformation they are taught in schools on evolution.

If you have children, you need to be aware of what they are being taught in our educational institutions and the sinister atheistic agenda forced on them. Millions of our youth are being brainwashed into believing the atheist religion of evolution, resulting in society's moral decline.

Understandably, many people have never taken the time to research the claims behind the theory of evolution. After all, is this going to change your life? Well yes, it will.

In a short space of time you will finally discover how evolution is not only an unproven theory, but how modern true science is proving it to be totally impossible. The information contained in this book kills off the Big Bang and evolution theories while conclusively proving creation.

Chapter five is especially fascinating, you will never forget what you discover there. The answer to an evolution topic debated for centuries is answered using biological scientific evidence, "What came first, the Chicken or the Egg". The answer to this age-old conundrum provides irrefutable evidence evolution is impossible.

Yes, the basic everyday humble chicken and egg creates a major problem for evolutionists and kills off their theory of evolution.

I trust you will enjoy this journey of discovery uncovering the true facts as much as I did. You will be enlightened and finish with a clear understanding of the impossibility of evolution.

COLIN B. NOBLE

Your Journey of Discovery Begins

Before we embark on any journey, we need to know where we are going. If we were taking a journey in a car, we would take time to research the best route to arrive at our destination using an application such as Google maps. This will help us figure out how long the journey will take, and where to stop for breaks or sightseeing.

The route might include a choice of taking nice wide freeways, making driving easy, or a way to discover interesting places with beautiful lakes or snow-capped mountains which may involve having to navigate narrow winding unsealed roads through forests or up steep mountain trails, requiring more concentration to avoid going off the road.

When we finally arrive at our destination, we appreciate that the journey was worth it, as it has enlightened us on something we never fully understood since we previously only read vague descriptions of our destination.

This journey and experiences opened our eyes to a new perspective, and we now have more knowledge.

Years ago, when living in New Zealand, our family embarked on an adventurous journey to one of the most stunning lakes in the world, Lake Waikaremoana. I hired a large motor home, unaware of the type of roads we were about to navigate in the days before Google maps.

There were no freeways on this journey, with the adventure beginning shortly after turning off the main highway. Our route was now taking us along unsealed rough and narrow winding roads. On a scale of 1 -10 on the worst road scale, this was easily a 9. It didn't take long to realize a large motor home was not the best type of vehicle for this terrain, especially when arriving at sections of single lane switchback corners perched alongside the side of cliffs that prohibited you from seeing the ground below. Thankfully, it was a remote area without much traffic. If I had met vehicles coming the opposite way on some of those blind single lane corners, it would have added to the fun by exponentially increasing the danger factor.

After many hours of slow dusty, dangerous mountain goat trails, the family questioning if our insurances were paid up, and wondering whose brilliant idea it was to hire a vehicle of this size, we turned a corner when suddenly before us unfolded a stunning, breathtaking view of one of the most panoramic lakes in the world. Wow, what a sight. The lake was surrounded by a giant podocarp rainforest, remote beaches, spectacular bush walks, and waterfalls. It featured bird life you don't see elsewhere, no pollution, pure clean air, and a night sky so clear you feel as if you can reach out and touch the stars. No houses, just a campground.

It doesn't get much better if you want to see the best of nature.

As we got closer to the lake, we came to a shape bend with a parking area overlooking the lake. It had just enough space for one vehicle and a perfect lunch and photo stop. I discovered I could reverse the motor home right up to the edge of the cliff overlooking the lake. That gave the family some anxious moments, especially when the view of the road under the rear of the vehicle disappeared. I also got even for my son's ribbing about whose smart decision it was to hire a motorhome.

As we sat at the rear dining table gazing out the window, you couldn't see any ground, just the incredible panoramic views of the lake. Best restaurant view in the world.

As we ate lunch and absorbed the magnificent views, there were no more snide comments about the roads we had to navigate, and the concentration needed making sure we didn't go over a cliff. All those were forgotten as everyone realized the destination was fully worth the journey.

You are about to embark on a journey, not in a vehicle, but rather in this book.

The benefits achieved from any journey depend on the level of planning by the organizer and driver, that's me. And how the passengers interact with what they discovered, that's you.

I like journeys to be fun and have endeavored to turn what could be a long tedious journey into a fun delightful journey by not making this a typical stuffy science publication.

Some chapters are freeways and easy to navigate. But to give you a rewarding adventure there are also some chapters with winding unsealed tracks; however, these are where you will find the journey's rewards.

Let's start our journey by firstly studying the road map to understand the meaning of the Big Bang and evolution theories, along with what they claim. We begin with the dictionary definition of The Big Bang:

big•bang n. The beginning of space, time, matter, energy, and of the expansion of the universe according to the Big Bang **theory***. [The American Heritage Dictionary]*

The Big Bang theory is a cosmological **model or theory** for how the observable universe was first formed and its subsequent evolution into what exists today.

Evolution Scientists claim a big explosion arising from nothing created the entire universe as all the matter we see went from being compressed into something smaller than a subatomic particle to expanding or rolling out from the point of the bang. This started everything, including time. Chapter three explains this theory in detail.

Common-sense logic, a phrase I will frequently use, tells us there couldn't be actual physical proof of the Big Bang and the start of life as no one was around to witness it happening. Also, the results of the Big Bang have never been replicated in the same environment that existed on day one to prove it could have happened, as no one knows what the atmospheric environment was when the bang occurred. No samples of the gas composition, before or after the alleged Big Bang, exist, and there are no samples of the first single cell lifeforms they claim every specie evolved from. Scientists only assume what was present for the bang to happen.

It's important to understand The Big Bang isn't fact, it's only a theory invented by scientists trying to explain how the universe formed without a Creator. There is no scientific evidence it happened, it's not proven, it's only theory. Even evolution scientists, when pushed, admit they have no direct evidence of it, yet they continue talking about it as if its proven science.

Evolution is another scientist invented theory starting after the Big Bang.

Ev•o•lu•tion n. 1. A gradual process in which something changes into a different and usually more complex or better form. .. . 2. a. The process of developing. b. Gradual development. 3. Biology a. The theory that groups of organisms change with passage of time, mainly as a result of natural selection, so that descendants differ morphologically and physiologically from their ancestors. b. The historical development of a related group of organisms; phylogeny. . . [The American Heritage Dictionary]

Notice that it describes evolution as *'a gradual process in which something changes into a different and more complex or better form'*. It then explains evolution is also only a 'theory'. There is no scientific evidence of the single cell evolutionists claim life started from after the Big Bang evolving into a blob of something, then into a fish, amphibian, insect, animal, or whatever their latest theory is. It's important to understand it's just a theory invented by scientists.

Evolutionists claim everything is in a continual state of change as they continue evolving into a better and more complex life form; that's their theory. Despite what evolution scientists claim, the fact is evolution is also unproven, with many monumental holes and no historical scientific biological examples. Scientists produce theoretical drawings of the envisaged stages of a species' evolution but this more than conjecture, not evidence. It's just their attempt at trying to support their theories.

The foundation of their evolution theory is that species developed over millions of years, with each generation improving slightly and developing into more complex life forms while those traits that are undesirable are weeded out. They claim evolution strengthens species, making them more resilient. This is highly flawed as historically a huge number of species have become extinct.

If evolution was true, statistics wouldn't show specie extinction rates continuing to increase. You will discover later extinction statistics that prove species don't become more resilient.

Evolution scientist's theories have so many flaws they are forced to continually invent new ideas which, if plausible, also becomes a 'theory'. This is the term scientists use for their unproven ideas, thoughts and assumptions as they endeavor to explain how the universe, earth and everything in it was created.

They then claim their theories are proven science, and that's how the evolutionists' publicity machine has brainwashed society into believing it.

Frequently on TV programs you hear claims from people with

limited understanding of the science behind Big Bang dating stating that their land or a few strange rocks are 300 million years old or some other number. It's so commonplace that the public innocently accepts it as fact when it's not. Dating rocks is very costly and complex, making it highly probable they arrived at their number by simply plucking a figure out of thin air that sounds good. Later in the book you will discover how unreliable dating methods are, further eroding the evolutionist's credibility.

Another alarm bell is evolutionist's unlimited number of theories on the precise mechanism by which life-forms changed and evolved from one type to another. One such theory is a fish grew arms, legs and lungs, then jumped out of the water and became a land-based animal. There is no scientific evidence of this ever occurring. Instead, it's actually science fiction, as you will discover.

Evolution scientists have no hard evidence of their wild claims, and this is where they play tricks to deceive the masses. You will hear them recite long and confusing theoretical arguments using scenarios they portray as being factual. They even throw in verifiable stories of a shellfish fossil found in a rock but then claim it's billions of years old, thus claiming this provides evidence to support their theory. While the fossil is real, it is not evidence of evolution as the dating method is highly flawed.

They also use an array of typical catch phrases such as, 'we believe' or 'we understand' or 'the science leads us to believe'.

What these phrases really mean is, we don't know and have no evidence, but a group of us 'think' it could be possible, maybe.

Their theories hold the same weight as gathering a group of people together who think there could be Martians.

This group then invents a plausible theory and then after telling the world they are all experts on Martians, they claim these beings exist because 'we understand' or 'we believe' Martians came from a distant galaxy and formed life on earth. The science behind that statement is as credible as many of the evolutionist's claims.

FACT: Evolution is a theory, and after reading this book you will discover it's impossible for it to be true.

As evolution is only a theory, we have to ask, why do around 55% of US adults believe in various forms of evolution; the theory that the universe, earth and all living creatures evolved over billions of years. Evolution scientists and their spin-doctors have conned the public into believing their unproven claims. They have also corrupted the educational system with science teachers who promulgate these unproven theories as proven science to their students, who innocently believe their science teachers. Some institutions take a very aggressive stance with teachers who don't believe in evolution and target students who believe in creation. Why are they so aggressive, for science is supposed to be the study of knowledge and seeking for the truth? They have an agenda. Yes, there is an agenda and reason for misleading students.

In a July 26, 2019 Gallup poll report:

22% Believe Humans evolved, and God had no part in the process.

This means 22% of Americans believe life only came from the 'Big Bang'.

33% Believe Humans evolved with God guiding the process.

40% Believe God created Humans in the present form and evolution played no part.

Based on these numbers, 22% have no belief in God and are probably atheists, so evolution conveniently fits with their denial of a Creator.

Those only believing in evolution and the Big Bang have risen from 9% in 1983 to 22% today. This is attributed to the biased educational curriculum that only teaches the flawed and unproven theory of evolution while quashing any dissent, as many science educators are atheists.

Of those polled, 33% of those with college degree and 16% with no college degree believed in evolution with no involvement from God, reflecting the strong evolution bias taught in the education system.

With the education system being highjacked by liberal atheist evolutionary academics who don't want youth hearing differing viewpoints, the above statistic reflects the influence they are having on misleading students. In a subsequent chapter, I expose their agenda and consequences.

A disturbing video on YouTube shows Richard Dawkins, an evolutionary biologist and leading evolution proselytizer and proclaimed atheist, proclaiming in a UK school science class that his sole purpose is to convince children to forget the traditional beliefs taught by their parents, who may be of the Christian or Muslim faith. Dawkins endeavors to convince students not to believe in God and that evolution is true. Misleadingly, one of his most well-known tactics is to quote information, which is only a theory, and then claim it to be fact. As part of his indoctrination, he takes them to a local beach where they look for evidence of evolution. All he can come up with is a rock with an embedded shell encrustation, claiming this is evidence of evolution.

No scientific proof, only his opinion.

https://www.youtube.com/watch?v=jNhtbmXzIaM

It is concerning that a school would allow a self-confessed atheist into the classroom for the sole purpose of undermining what parents taught their children. This shows a total lack of respect for the children's parents and an example of the extreme measures hard core atheist evolutionists such as Richard Dawkins are willing to go to in promoting their atheism.

Dawkins is on record claiming if parents influence their children's beliefs, it's a form of child abuse. Remember, this man is a militant atheist who doesn't hesitate to mock God and religion any chance he gets. These are the type of people educators are allowing to teach and influence your children.

More on Dawkins and the atheist movement's tactics in promoting their religion of evolution in following chapters.

Methods of Evolution

They base their theory of evolution upon two processes, mutation and natural selection, which result in billions of incremental changes in organisms over a vast time period. The assumption is that these various infinitesimal changes randomly and amazingly fall into a perfect pattern as if an intelligent force were guiding each step. Basic common-sense logic and reasoning asks how this could happen. We know from the complexity of the human body that it's an incredible design, and so complex as to be impossible for it to develop with no initial design, instead occurring following billions of infinitesimal uncoordinated random mutations over millions of years. I'm not a gambler, but consider the incredible odds of every little step in the evolution of absolutely everything coming together in a perfect order over millions of years with no one controlling it, and no species without this order, such as having an arm coming out of the forehead.

Mutation

The process of a living organism, animals, humans, fish, birds, or any living thing changing the DNA/genes very slightly in its offspring, resulting in each subsequent generation undergoing infinitesimal changes. This means their DNA, which are the building blocks of every living being, would need to continuously only change for the better, and that's where their theory has a huge credibility problem. How that could ever happen is science fiction as no specie has influence on the changes of its offspring's DNA. Evolutionists obviously can't explain how this actually takes place as DNA is extremely intricate.

DNA code is more complex than the most advanced computer coding. No human could possibly invent it, however we are expected to believe it all just fell together by random mutations. That's biologically impossible, as you will read in chapter six.

Remember, they claim mutations improve the specie, so it is vital for this to occur, as this is an essential element of their theory. If just one minuscule mutation went wrong in the DNA, the specie would be extinct. Their claim of only positive traits is fantasy and it hasn't worked out for them as there are thousands of extinct species and the number is rising rapidly. If just one random mutation went the wrong way in early humans, we would have been extinct millions of years ago.

Now multiply that by the billions of mutations they claim took place to arrive at a human and you can see how impossible that is.

Their claim is crazy, it's basically saying if you had a block of land with trees on it, then over a few million years these trees would evolve into a house.

Evolutionists always like to think mankind is in control of their destiny, that's the one reason why they refuse to accept intelligent design. Therefore, if we follow their science fiction theory and if it was true, every parent could change the DNA coding in their children so they would be born with modified DNA to make them smarter, stronger, a top golfer, tennis ace, or whatever the parent desires.

We know we can't do that; the power of positive thinking doesn't go that far.

Whilst the theory sounds appealing and your children might like to think they are evolving to be more intelligent and smarter than you, they have a problem proving it as there is no scientific evidence this is happening. Sorry kids.

But who are we to question the fantasy of Evolution's invisible magical force? Sounds like they may have been watching too many Star Wars movies. May the DNA force be with them.

Natural selection

This means the weaker of the specie dies off through illness or being eaten by larger animals, while the stronger of the specie survives, passing those genes to their offspring.

Once again, you're probably asking yourself a very logical and common-sense question. If the stronger of a specie passes its genes to its offspring with a slight mutation, making them even better equipped to overcome obstacles that may arise in its environment, which is the basis of evolution, then no specie should ever die off. That's their claim, yet to us they contradict themselves, it's illogical.

But this makes perfect sense to the evolutionists as they claim somehow this magical, invisible, mysterious, non-observable force with no brain, lifeform or intelligence miraculously determines which specie dies out and which ones survive. It has no plan; it doesn't physically or spiritually exist, it's a non-existent nothing, yet for billions of years it has designed the universe, earth, all of nature, animal life, humans, food sources, absolutely everything.

When evolutionists are asked to answer these basic questions or explain the many contradictions, they are never able to provide an intelligent explanation. Their brains apparently haven't evolved enough to answer this fundamental question, so they use their stock reply out of the evolutionist's science fiction manual. It's their usual lame escape clause when cornered. Here it is from a typical evolutionist website:

It is a process that over billions of years, gradually selects organisms that are better adapted to their environment, and in this way takes advantage of random mutations continuously changing life and makes all living organisms in existence be the way they are.

Evolution is not a finished event, of which we, humans, are the final product. Rather, it is a continuing process which has been changing and forming life on earth for billions of years and continues to do so for as long as organisms are being born, dying, and competing for what they need to survive and reproduce.

That is just dribble. Where is the evidence? Evolution hasn't been observed or scientifically proven, it's all just an invented theory.

Note how despite evolution being an unobservable entity or force, they personalize 'evolution' to conceptualize it collectively as a thinking being or force that selects those species best suited to survive. Even in their subconscious they know how impossible their theory of the mystical evolution force without intelligence is.

An evolution supporter once wrote, "*The creation model (God created the universe) is difficult to accept because it assumes an external agent that has not been observed.*"

A classic example of an evolutionist's hypocrisy is claiming he can't believe in creation because he hasn't personally observed God.

Hello, does he realize what he just said, evolution hasn't been observed either. They have never seen evolution in action; have not seen or witnessed actual progressive changes physically taking place before them from one specie to another. No one has witnessed a fish changing to a land mammal as they claim or viewed the vast mass of gasses forming a living cell, which they claim was how life began.

Evolutionists can't physically observe evolution in action or see it actually happening. They haven't observed it in the same context they refer to observing God, yet they conveniently believe evolution theories and make unfounded statements claiming it's true just because some scientists invented a theory.

Whilst they mock creationists for believing in an invisible Creator, they can't physically see the so-called invisible force of evolution. They claim one can see evolution by looking in the mirror as we are all the product of evolution. This isn't evidence, as they cannot show physical evidence of all the various mutational stages of a human. Their whole proof is based on unsubstantiated drawings of what they assume happened.

Evolutionists can't call up the force of evolution from wherever it hangs out. Evolutionists are simply putting their faith in the unproven theories of mortal infallible scientists.

If no one has observed the force of evolution and it's only a theory with no direct evidence, why do people believe it?

There are some people who, despite the overwhelming evidence supporting creation, simply don't want to believe in a Creator and supreme intelligent being. For these individuals, evolution is a convenient alternative to suit their belief in atheism for it removes any accountability for doing right.

Has science proven evolution true?

What is science? Sounds like a silly question as everyone knows its dictionary definition:

the intellectual and practical activity encompassing the systematic study of the structure and behavior of the physical and natural world through observation and experiment.

If that's true science, then what can be claimed as science? If I invent a chemical formula claiming it will cure a disease, is that science? No, it's only a theory. I claim it will cure a disease, however, how do you know it will do what I claim as it's just my words and a bunch of numbers on paper. Even if other scientists agree that in theory it will work, it's still only a theory. Would you simply trust a theory and my claim it will work if you wanted to use it on yourself or a loved one to cure some illness or disease? No, you wouldn't because the side effects might kill you. You would be crazy to blindly believe my words, even if I was a scientist.

You would rightly demand to see evidence it works. You would want to see my research proving it was comprehensively tested according to established science protocols, as you want assurance it works not just once in artificial laboratory conditions, but many times on real people so any potential side effects can be identified that might render the claims of the alleged cure and benefits unreliable, and therefore a fake cure.

Evolution scientists rashly claim evolution is true and a proven science, however it has never passed established science protocols, and herein lies the monumental deception and con of evolution.

An article published by Pew Forum on Feb 6, 2019 titled *'Darwin in America'* states,

Most biologists and other scientists contend that evolutionary **theory** *convincingly explains the origins and development of life on Earth. Moreover, they say,* **a scientific theory is not a hunch or a guess but is instead an established explanation for a natural phenomenon, like gravity, that has repeatedly been tested and refined through observation and experimentation.*

So, if evolution is as established in the scientific community as the theory of gravity, why are people still arguing about it more than a century and a half after Darwin proposed it?

The answer lies, in large part, in the theological implications of evolutionary thinking. For many religious people, the Darwinian view of life–a panorama of brutal struggle and constant change–conflicts with both the biblical creation story and the Judeo-Christian concept of an active, loving God who intervenes in human events.

I doubt anyone on planet earth would dispute gravity exists. While we can't see it, we experience it in various forms because it's an ever-present force, it's not a process. The same is true regarding the air we breathe, we can't see oxygen but we sure know it's there, it's an ever-present fact of life, we experience it first-hand; we know it's true, no oxygen and we die. Lock yourself in an airtight trunk to conduct an experiment on the theory that humans need oxygen to live. You will find out very quickly if oxygen is real or not. You will experience the lack of it first-hand and your experiment will prove the theory humans need oxygen to survive.

Evolution scientists, in their desperate attempt to prove their theories true, want to bypass the foundations of the scientific method because they don't have quantifiable evidence, however, they want their unproven theories to be accepted as science on the basis of their say-so.

The definition of "theory" means *a supposition explaining something*. It is not fact; it's not experienced first-hand like gravity or oxygen, and to use gravity as a comparison is an act of desperation.

For a theory to become science it must pass the established criterion of tests, for these are the foundations and principles of true science. Evolution scientists belittle the long established and respected scientific processes relied on to prove something is true by testing it under reliable and trusted protocols to see if their theory can be duplicated. Evolution scientists want us to accept their unproven theories as true science just because they want to build their religion of atheism.

They demean and belittle the truth of science by claiming a theory is science, and if they can get a bunch of their science cronies to agree with it, then in their world this makes it true!

The same people also claim Charles Darwin's theory was groundbreaking, an incorrect statement on its face. Darwin wasn't the first person to come up with the theory of evolution, instead he plagiarized a lot of his writing. Evolutionists need to read all of Darwin's books, along with the writings of earlier evolutionists. Just because the seed of a theory has been around for 160 years doesn't prove it's true.

In Darwin's era, he had many critics including well-known and respected biologists who strongly disputed his writings. Darwin's known obsession for acceptance of his theories by his peers is what drove him to write books.

Despite his best efforts, this acceptance eluded him.

Evolution has **NOT** passed the longstanding and accepted protocols of science to prove something is true.

Science is built on the foundation of the assumed purity of the scientific method.

The five basic steps in a scientific method are:

1. Make an observation, identify and define the problem

2. Ask questions and gather data

3. Form a hypothesis or testable explanation

4. Make a prediction based on the hypothesis

5. Conduct experiments, test the prediction, prove the theory with testable results that can be duplicated.

For the evolutionary process to pass the protocols of true science, it must be observable and tested.

The start of the universe was never witnessed. How did they measure the various gases and their composition?

What's the actual formula of the mixture of gases and their various quantities, and have they successfully replicated it in the same environment to prove it works. NO, they haven't. They are only guessing as they don't have the formula. It's not tested, and scientists have not recreated the 'Big Bang' to prove their theory is scientifically correct.

To claim the Big Bang and evolution is science is a fraud, a lie and a con.

To put the con in context, would you believe me if I told you that one day, I gathered a bunch of gasses. I don't know what they were or their quantity, but I released them into the atmosphere and surprisingly, they somehow found each other, stuck together, made a reaction and eventually formed a living cell. I know you wouldn't believe me and would probably tell my wife not to let me out of the house again. But this is what evolutionists tell us, not in plain, simple words, instead they use confusing abstract words and jargon. However, if you shake out the big words, this is exactly what they are claiming and want us to believe.

Unless there is a 14 billion-year-old scientist who observed the process and all the various stages of a single living cell evolving into a living specie with supporting evidence that can be tested and reproduced, it's NOT science. It's only a science fiction theory and always will be, as it's impossible to prove.

Developing theories and making endless unproven assumptions are not science.

To arrive at a scientific conclusion, it must be tested and repeated. If a discovery can't be repeated, it's not real and only exists in the person's imagination. The testing is what makes it science. I've never read where a scientist successfully recreated the big bang and ended up with a new living cell.

We must always remember they have **NO EXPERIMENTAL PROOF;** scientists can't prove the Big Bang or formation of the first living cell evolving into other living forms that ended up becoming today's species, including you and me.

The theory of evolution isn't science, it's only an idea and the crazy part is it's taught in our educational establishments as fact.

Shame on the establishment for conning their students.

Buckle up your seat belt. You are about to unearth many amazing discoveries as you navigate subsequent chapters. You will find interesting and awesome facts around every corner, so don't miss a turn in the road. Discoveries are to be made of undeniable biological scientific facts and examples, proving without a doubt the theory of evolution is a con.

When you arrive at the end of your journey, you will be much wiser and enlightened as to why the Big Bang and evolution is impossible and the greatest scientific con of all time. It may even change your life.

Where Did the Theory of Evolution Come From?

The theory of evolution is a recent invention. As a student in the late 1960s and early 70s I never heard it mentioned.

In 1925 a USA high school teacher was arrested, tried and convicted for the crime of teaching evolution in his class. In the late 60s this law was challenged, and it is no longer illegal to teach evolution in schools. Instead, courts have now ruled evolution is the only theory regarding creation that can be taught.

The atheist movement behind promoting evolution credits its discovery to Charles Darwin in the 1800s. Evolutionists claim Darwin's findings scientifically proved evolution, but this is a gross exaggeration. Not only did Darwin not scientifically prove evolution; he didn't even come close. In fact, he never used the term 'evolution'.

Despite this, Darwin receives accolades from evolutionists.

They revere his writing's and opinions, using them as the basis of their belief in the theory of evolution. Darwin is the god of evolutionists and atheists. However, Darwin's life and writings contains numerous discrepancies and misrepresentations of their claims.

This chapter exposes the real Darwin and the evolutionist's lies surrounding his discoveries.

When we study Darwin's writings in detail, we quickly discover that their god definitely did not invent evolution or prove it to be true. Using him as the fundamental cornerstone of their belief system severely damages their claims from the outset.

Scholars, from ancient Greek philosophers on, have long inferred that similar species descended from a common ancestor. The word 'evolution' first appeared in the English language in 1647 where it was used in a nonbiological connection before becoming widely used as the term for species progression from simpler beginnings.

The term Darwin most often used to refer to what is now called biological evolution was 'descent with modification.'

Many people who claim to be evolutionists have never read Charles Darwin's book, *On the Origin of Species by Means of Natural Selection*, published in 1859, or any of his other books. Were they to do so, they would become very disillusioned. It is fascinating how evolutionists freely declare Darwin discovered evolution and bestow upon him the title of 'father of evolution' yet have never taken the time to read the bible of their god.

Darwin's motivation and agenda were to disprove the existence of a Creator, which he never did. Were he alive today, he would be very happy to know how his mere observations and thoughts were being used to con the world into believing the theory of evolution. His books contain a litany of unproven assumptions and personal opinions.

First, some background on Darwin.

Charles Darwin, 1809 -1882 was one of six children from a 19th century wealthy British family. He enjoyed a very privileged life, with financial backing from his father enabling him to attend university.

Darwin's grandfather, Erasmus Darwin was a physician with an interest in natural science and philosophy. Erasmus published a book on the laws of organic life which formed the basis of the early ideas that later became Charles Darwin's theories.

Charles' mother died when he was eight and shortly thereafter his father sent him to a boarding school.

He didn't like school and wasn't a good student. He preferred catching beetles or hunting, compelling his father to decide Charles would follow the family tradition and work as a medical assistant in his medical practice. Young Charles hated it, preferring to enjoy the lackadaisical life of a wealthy privileged son.

Whilst at university studying medicine, Charles discovered an affinity for researching animals and organisms, leading to an interest in taxidermy.

Convincing his father medicine wasn't for him, Erasmus decided Charles would make a good clergyman in the Church of England and sent him to Christ College. During this era parents decided their children's careers, who then obeyed; however young Darwin preferred a life of leisure and partying.

After graduating in 1831, Charles received an invitation to join the crew on HMS Beagle for a mission to survey the coasts of South America and bring back useful information for the Royal Navy. Darwin's role was to collect specimens and observe.

Darwin was not a sailor and suffered severe sea sickness which confined him to his cabin for much of the voyage, making him wish he had never left the comforts of home. Being the 1800s, he couldn't jump off at the next port and fly back home. Instead, he remained on the Beagle for five years, living off a very simple diet while endeavoring to overcome his sea sickness.

As the Beagle spent its time charting the coast, Darwin would go ashore whenever possible to collect samples and observe local geography while compiling copious notes on his observations.

Their first stop was Madeira Island, but they were unable to land because of bad weather. Sailing on to Tenerife Island, they couldn't land due to a cholera outbreak. They upped anchor and sailed to the Cape Verde Islands and Santiago.

While there, Darwin explored the Spanish ruins, but he spent most of his time on the beaches. Cuttlefish fascinated him as they could change colors, which he thought was his discovery alone. Later in England he found this to be common knowledge.

A white line on a cliff face around 10 meters above the sea attracted Darwin's attention.

On closer examination he found the line was a layer of shells embedded in the rock. He pondered how they got there, being so far from the water. This discovery was the beginning of his observations that led him to guess how this may have happened. As he did so many times before, Charles observed and made assumptions.

He read books on how volcanic activity caused land masses to slowly rise and oceans fall; at least he assumed this was the case, reasoning the slow rising of the land prevented the shells being crushed. And so began the birth of his theories contrived from observations.

Arriving at Buenos Aires, the ship stopped for mail and supplies. While there, Darwin dispatched his collection of samples to England, consisting of rocks, tropical plants, beetles, some animals preserved in alcohol, along with a large selection of marine specimens.

When the Beagle arrived back in London in 1836, Darwin's wealthy father continued to support Charles, enabling him to focus all his time on his amusements. With no necessity to work, Charles had all the time he wanted to contemplate and develop his various observations and theories of what he had seen during his time in the South Pacific.

During the course of his 5-year adventure on the Beagle, Charles managed to collect around 4,000 specimens that required cataloguing. This would have been too much like hard work for Darwin and detracted from his social life, so he arranged for others to do the tedious task.

Charles ended up at Cambridge where he devoted his time to studies and preparing a variety of papers, one on the theory that the landmass of Chile was rising. This was not based on historical evidence as he made just one visit to the area on an old sailing ship. It was just his idea, a theory in his mind. So many of evolution's theories are arrived at in this manner, following Darwin's tradition.

An ornithologist observed some bird samples Darwin bought back which he called 'strange birds'. Darwin thought they were types of bird-like blackbirds and finches, when in fact they were all finches. While collecting the birds, Darwin failed to recognize them as finches, yet his Finch Theory is a cornerstone of evolution.

Darwin's Finch theory

One of the stops along Darwin's 1835 Beagle adventure was the Galapagos Islands in the Pacific Ocean, some 1000 km off Ecuador. The Galapagos Islands are an archipelago of 13 major islands and more than a hundred smaller islands. Their location provided an isolated terrain containing a diversity of plants and animal species, some found nowhere else, hence its uniqueness.

Whilst observing bird life he noticed a group of finches that appeared to be related to each other in the structure of their beaks, short tails, and bodily form and plumage.

Curious how the finches had various shaped beaks, Darwin assumed they may have originally come from a single specie which then developed different sized and shaped beaks to adapt to the food sources. What Darwin called the ground-finch (Geospiza) had a thick beak adapted to feeding on a variety of crunchy seeds and arthropods, whereas the warbler finch (Certhidea olivacea) developed a slender, pointy bill to catch tasty insects hiding between the foliage. The woodpecker finch (Camarhynchus pallidus) used twigs or cactus spines to pry arthropods out of tree holes.

On the basis of these observations, Darwin's theory arose. He had no historical evidence to back up his beliefs, instead Darwin just assumed they rapidly evolved to survive in the harsh environment. Obviously, there is no observable evidence of this whatsoever, Darwin merely observed and immediately made an assumption. With no long-term generational observations, Darwin jumped to a conclusion so he could invent a theory.

Evolutionists love Darwin's finch observations and irresponsibly use them as evidence of evolution.

A logical question one would ask is, what historical evidence exists that these birds actually changed over the years. Can they show any examples of the finches from 1 million years ago proving the evolutionary changes? No, they can't? There is no historical evidence proving their claim.

Creationists claim God created the Finches on the Galapagos Islands with those differences so they could survive, and evolutionists cannot disprove that statement as they have no evidence of the evolutionary stages of those finches.

Once again, Darwin's beliefs are just a theory, yet they refer to it as fact.

Look at the facts and you be the judge. Darwin only visited the Galapagos once, where he saw a few finches with slight beak differences before getting back on the boat. He never saw the finches again, but 160 years later evolution scientists still claim this brief excursion provides ironclad evidence of evolution. Any common-sense person would agree that not only is this not evidence, it's illogical.

Another interesting event is currently playing out on Galapagos which challenges the evolutionist's theory that the finches changed their beak shapes as a means of survival. One of their key elements of using the Galapagos Islands for their 'laboratory' of evolution is what they believe is its 'closed ecosystem'. However, into this pristine environment there has entered a literal 'fly in the ointment'.

The *Philornis downsi* fly has immigrated to Galapagos. This insect jumped onto a cargo ship in the 1960s, where they purchased a one-way ticket and took up residence on Galapagos.

This small creature is now creating quite an interesting situation. 'Earthwatch' volunteers are now having to monitor the survival of Darwin's finches, concerned the fly is having a devastating effect on the finch population to where they could soon become extinct.

The reason for this is the adult fly eats fruit and then lays its eggs in the finch's nests. Once the eggs hatch, the larvae feed on the blood and tissue of the nestlings. Since 2000, researchers have observed an alarmingly high nestling mortality rate, with anywhere from 30 to 98 percent of chicks dying each year.

Since Darwin claimed the finches evolved to become more adept at survival, researchers have been diligently watching and waiting to see physical evolutionary changes in the finches to combat the effect of Philornis. Don't know how that will work out for them as they teach evolutionary changes take millions of years. To date they have reported no changes.

According to the theory of evolution, a species evolves over millions of years, making infinitesimal changes every generation as required to survive. One of the fundamental arguments from Darwin's finch observations was how they quickly changed the shapes of their beaks to survive, inferring it doesn't take millions of years. Do you see how they quickly change the claims to suit the situation?

Evolutionists' faith in proving their theory is currently being tested in the same laboratory where it began. Will the finches evolve to combat the *Philornis* parasite, as Darwin 'observed', or will they become extinct?

The result is in, and they failed the test. Evolutionists proved they didn't actually believe their theories and folded, so much for their confidence in evolution as taught by their god Darwin. They don't practice what they preach by doing the unpardonable, interfering with the all-knowing, always right force of evolution by introducing control measures for the *Philornis* parasite. And now these hypocrites claim their work saved Darwin's finches from extinction. Don't they preach evolution is supposed to do that, not man, and has done so for billions of years.

A sad day for evolutionists as they lost complete faith in evolution, discovering it doesn't happen as their theories claim. However, it's a great day for the finches who may now survive, no thanks to evolution.

Isn't it ironic that the evolutionists didn't even have enough faith in their own theory to let it play out, thus validating it for the whole world? If they believed evolution was true, they would have sat back, knowing evolution's magical force would work out for the betterment of nature, even if that meant the finches became extinct. These people obviously have no faith in their belief of natural selection.

Fact is, if evolutionists hadn't interfered the finches would be extinct in a very short time, thereby proving Darwin's theory wrong; specie can't evolve to survive. I wonder if there are any evolutionists willing to be honest enough to re-write their stories on Darwin's finch theory to include how they were on the brink of extinction, but they couldn't evolve fast enough to survive. Instead, it was only human intervention that saved them, an 'intelligent designer', proving Darwin was wrong.

This is just one very simple example of the fundamental flaws of evolution. This one specie would have been extinct in a very short time span, it did not evolve to survive.

Consider all the thousands of extinct species of just the last 300 years. If evolution was true, these species would have evolved to survive. You can't make the claim that only a selection of specie evolves whereas others don't know how and become extinct, how does that work.

Evolutionists claim the magical force of natural selection decides, causing the weaker species to die off. That doesn't work either as it contradicts their argument that species are capable of evolving to survive. If that's true, why do any die off.

We see with Darwin's finches that over a short time-span, 10-20 years at a guess, they would be extinct if left alone, proving Darwin's theory wrong.

Evolutionists claim this invisible magical force of evolution and natural selection has the power to control which species survive earth's food shortages, storms, disease and pestilence. It's an unidentifiable mystical force with no intellect that somehow allows these mechanisms to cull the species so only the strongest survive and earth doesn't become overpopulated. Obviously, they can't prove this claim.

It is no surprise to learn that many climate change advocates are also evolutionists. If evolution is true and natural events are part of the evolutionary process, then why are they involved in climate change activism and wanting to save the planet?

Why does the planet need them to save it, as this supposed warming is part of the very evolutionary process they preach? According to them, this magical force of evolution has been operating in perfect equilibrium for billions of years since the beginning of all creation, so why are they trying to interfere with it. Perhaps the force of evolution wants the planet to heat. Evolutionists teach the force of evolution knows best and has worked just fine for 14 billion years, so what gives us the right to interfere.

They tell us earth today is the result of a series of random uncontrolled mutations that occurred over billions of years. Evolution created the universe, earth with all the living creatures, the human body, and absolutely everything in perfect order and design. That's what evolution teaches, so why are they interfering and worrying about climate change. Do you see the irony here?

If what they preach is true then evolution is more than capable of dealing with climate change, just as it has for billions of years? They have no confidence in the reality of what they believe and teach.

Back to Darwin's life.

After Cambridge, Darwin began mixing with other like-minded individuals who fostered his ideas. During this time, Darwin rewrote his notes from the Beagle journey and edited the reports on his collections. At age 30 he developed a health issue he would struggle with for the rest of his life.

When he met his future wife, he deliberated for some time if he should marry her, writing a list of Yes's and another of 'No's'. The yes's won out and he married Emma in 1839, a third cousin and devout Christian. Initially, she refused to marry him, fearing she would not see him again after death if he wasn't a Christian. They came to a compromise with Darwin agreeing to have faith in Jesus, who in the Christian faith is the Son of God.

In Darwin's notes, he commented; *"if it Please God"*. Interesting comment for a person who evolution atheists tell us was the father of their atheistic evolution religion which denies God exists. Darwin also referred to the *'act of creation'* in his book, On the Origin of Species, a fact that evolutionists conveniently always forget to mention. Perhaps Darwin actually believed in God and creation; that thought should give evolutionists some sleepless nights.

As a wedding present, Darwin's father gave him a significant amount of money. Investing it wisely, Darwin never had to work, which allowed him to devote all his time and energy to writing his books, with Emma editing them.

Evolutionists never mention that a devout Christian edited the very same books the atheists use as their bible.

Here is a very interesting fact, I'm not aware of any of Darwin's writings where he denies the existence of God, rather he commented that he couldn't explain the complexity of many species, including the human body, which couldn't have simply evolved. There is plenty of evidence Darwin believed in a Creator.

Darwin's health continued to deteriorate. He suffered from vomiting, stomach pains, boils, headaches, and heart palpitations, all of which restricted his activities. Travel made his condition worse, so he spent most of his time at home writing books with Emma nursing him.

In 1848, Darwin's father died. This utterly devastated him. Shortly thereafter his oldest daughter became ill with similar symptoms to his, and she died at the age of 10, which took a tremendous toll on him.

Charles spent eight years studying his collection of barnacles, desperate to find evidence for his ideas proving different barnacles all had a common origin and changed body parts to adapt to new environmental conditions to form the basis for his evolution theory.

How could a group of specimens taken over a short 5-year period prove evolution when evolutionists tell us it takes millions of years, yet they continue to spin the tale that Darwin proved evolution. This is nonsense. Darwin had no evidence of transmutation evolution, he couldn't show stages of changes over a long period, and all he could do is theorize and make assumptions.

Because barnacles shared a common structure with no discernible differences, he assumed they all came from one original specie and went on to change to suit their environment. No evidence, just another assumption.

He then moved on to studying seeds and plants.

While working on his flagship book, *On the Origin of Species*, a friend from Borneo sent Darwin his manuscript on natural selection, asking him to publish it as he was ill with Scarlet fever. This was the same topic Darwin was working on and his friend beat him on the completed manuscript. Despite his disappointment, he organized the printing, however it wasn't a success and was widely criticized by their peers.

Darwin finished his book, *On the Origin of Species*, initially printing 1,250 copies. Interestingly, he didn't use the term 'evolution' in the book, so for evolutionists to state Darwin discovered evolution is wrong in both fact and substance.

His health continued deteriorating to the stage where he no longer attended debates on his writings.

Many of Darwin's peers criticized his writings, fueling an obsession to write more of his observations in his continual striving for their acceptance, which was his driving force.

If any evolutionist hasn't read all of Darwin's books, especially the evolutionist's bible, *Origin of the Species* from cover to cover, but still refers to Darwin as having discovered evolution, they have deceived themselves and done themselves a great injustice. No logical rational person could ever objectively come to the conclusion that he discovered evolution. His writings are just thoughts, assumptions and theories taken only during his short lifetime with no scientific supporting evidence.

If you don't believe in evolution, don't waste your time reading his tedious writings, there are far more interesting topics to read.

Darwin's own words contradict evolutionists claims he discovered evolution; He readily admits his theories were nothing more than mere speculation:

"After five years work, I allowed myself to speculate on the subject".

And here is a revealing line from his book confirming he used unsubstantiated 'Statements',

"This Abstract, which I now publish, must necessarily be imperfect. I cannot here give references and authorities for my several statements".

Darwin's writings were built on assumptions, and this tradition continues today with evolution scientists. The entire foundation of evolution is built on unprovable assumptions.

Unless evolutionists can prove transmutation of specie, e.g. the various stages of change from a fish to a land mammal, which they claim happened, they cannot prove evolution. There is no evidence of this ever occurring. Even Darwin, the god of their evolution theory, knew this transmutation didn't exist. When Darwin states 'authorities' he means proof, that's how they said it back then. Evolutionists never quote this Darwin statement.

Here's' another Darwin truth evolutionists never quote and preferred didn't exist:

"In considering the Origin of Species, it is quite conceivable that a naturalist, reflecting on the mutual affinities of organic beings, on their embryological relations, their geographical distribution, geological succession, and other such facts, might come to the conclusion that each species had not been independently created, but had descended, like

varieties, from other species. Nevertheless, **such a conclusion, even if well founded, would be unsatisfactory, until it could be shown how the innumerable species inhabiting this world have been modified,** *so as to acquire that perfection of structure and coadaptation which most justly excites our admiration".*

Here are Darwin's own words admitting he couldn't prove evolution as there was no scientific proof the species inhabiting this world evolved over time. He had no evidence of transmutation specie, which the evolutionists still don't have and never will have.

It's what I call a bombshell confession from the father of evolution. He unabashedly admits he can't substantiate his theories with evidence. This is why, even in his day, Darwin's peers criticized his work, because he couldn't prove it.

Even with Darwin's confession of having no proof supporting his writings, many in the scientific world, academics and atheists think he is a god, even going so far as to make the most stupid claim he was a genius. If that's their interpretation of a genius, it explains their irrational claims evolution is true; they live in a realm of science fiction.

To summarize Darwin's book *Origin of the Species* in one sentence I would conclude:

"It's a book of one man's personal observations on the large variety seen in living species and his thoughts, assumptions and possible scenarios on how these variations came about."

One could compare it to the writings of a poet on a subject that interests him. Darwin's book merely documents his observations on diversity seen in species as he ponders the reason for it. He didn't provide evidence of evolution; he espoused a philosophy.

He pondered many things, such as why some animals breed well in the natural habitat but not in confinement, whereas others do.

He pondered how a rabbit could be bred prolifically under any circumstances and came to the conclusion that their change of environment hadn't affected their reproductive system. He simply pondered the reason why, never providing documented answers or supporting science.

Darwin's writings are littered with his frustrations of not being able to prove his theories, often using these phrases:

"If it could be shown; I should think; but who can say; I suspect; whatever the cause may be; as I believe; might make;"

I could add lots more, however I'm sure you get the point.

He thought domestic dogs descended from several wild species. He merely 'thought' this was so, then he states, *"In regard to sheep and goats I can form **no opinion**."*

In his writings Darwin rambles on with a never-ending tedious list of observations. He never provides any scientific findings to substantiate his thoughts.

No one denies there is variety in species, however there is no transmutation evidence of a specie detailing how it changed from one form to another. No one has photographed, recorded, documented, or proven how a species has changed over the million odd years or whatever pie-in-the-sky number evolutionists randomly throw around to make themselves sound like they actually know. No one denies changes in a species through selective breeding can happen, it's common in the racehorse industry as they try to get the fastest horse.

However, what we don't see is a horse naturally producing two rear kangaroo type legs to propel it faster. A change instigated by a horse thousands of years ago in a stable in Australia because he had an epiphany one day when he saw a kangaroo and said to himself, 'if only I had a couple of legs like that, I would win every race', then somehow his DNA/genes started the transmutation process. Genes, of course, don't have brains, so goodness knows how they remembered his idea to change the gene code and pass on the message to all subsequent generations of the horse's offspring to continue fathering the idea to develop kangaroo legs.

Yes, this is a simplistic analogy, however this is how silly the idea that a fish or frog can somehow become a dinosaur as espoused by evolutionists is.

Contrary to what evolutionists claim about Darwin's findings, his book is full of statements confirming he didn't prove transmutation of specie or evolution as he readily admits to *"unknown elements"*.

Here is another Darwin bombshell confession you never hear evolutionists mention. Darwin acknowledged the act of 'creation'.

*"...yet every naturalist knows vaguely what he means when he speaks of a species. Generally, the term includes the **unknown element of a distinct act of creation.**"*

(Darwin, *The Origin of Species*.)

Despite the claim Darwin discovered evolution, this is untrue as Darwin's own words prove. He asks, "how did this happen", these are the words of someone who hasn't made a discovery.

*"But the mere existence of individual variability and of some few well-marked varieties, though necessary as the foundation for the work, **helps***

us but little in understanding how species arise in nature. How have all those exquisite adaptations of one part of the organisation to another part, and to the conditions of life, and of one distinct organic being to another being, been perfected?"

"Again, ***it may be asked, how is it that varieties, which I have called incipient species, become ultimately converted into good and distinct species, which in most cases obviously differ from each other far more than do the varieties of the same species?"***

(Darwin, The Origin of Species.)

In a section outlining how some species rely on others for survival, Darwin admits he can't work it out and is ignorant of how all this works together, so much for the discovery of evolution.

"Probably in no single instance should we know what to do, so as to succeed. ***It will convince us of our ignorance on the mutual relations of all organic beings;"***

(Darwin, The Origin of Species.)

There is an overwhelming theme in Darwin's observations on the wonders of nature. He recognized how fragile the structure is. How disease, pestilence, and survival of species in times of significant weather events can change the balance of nature. Whilst observing these events and labelling them as events of natural selection, he fails to provide any insight on how they are ordered. According to Darwin they are simply random events; nature is a fragile balancing act and if events got out of control, it would wipe out entire species as they can't evolve to survive.

As I write this book, the Australian bush fires are having a devastating effect on Koala bear populations because they move too slowly to escape the fires. Large bush fires are a long part of Australia's history, they are not recent events created by global warming. Despite the fires being a normal event, the Koalas have not evolved in any way to outrun fires or develop fireproof fur.

Bees are another perfect example. If a disease or parasite wiped them out of existence, our food supplies would be in danger as bees are a main source of pollination. In recent years, we are seeing parasites affecting bees and they have no natural defense mechanism to fight it. The parasite Varroa mite can feed and live on adult honeybees, but they mainly feed and reproduce on larvae and pupae in the developing brood, causing malformation, weakening and transmitting many viruses. Eventually these parasites will kill off honeybee hives and nests.

If evolution was true, why haven't bees evolved to fight this threat to their existence? Why hasn't the mystical force of evolution reacted to save the specie? This has just happened in my lifetime, represented by just one grain of sand on all the beaches of the world compared to evolutionist's14-billion-year time scale. Think about it, if evolution was the only force at work in nature, over the many billions of years every living thing would have been extinct a long time ago. And that includes us as we would never have evolved to this point.

The USA has Varroa with bee populations in rapid decline. Local bees are mysteriously disappearing, they suspect something is attaching to their immune system and have had to import Australian bees to pollinate orchards as there is no other natural pollination available.

There are many other examples where parasites, pests, and diseases would have exterminated a specie if mankind had not intervened, and that's just in recent times as we have developed the necessary expertise and technology. Evolutionists claim mankind in its current form is around 200,000 years old, which is very young on the scale of 14 billion years. Before mankind existed, who intervened to save a specie, no one.

This is where the theory of evolution is seriously flawed. It tells us species evolve to survive, however there are so many recent examples where, if mankind hadn't intervened, these same species would be extinct. Proving if evolution was true, species would evolve at a pace to protect themselves. E.g. bees and Darwin's finches would have developed a protective mechanism to fight viruses. It just doesn't happen, and evolutionists can't give any examples that it does. History is full of extinct species, and new species are not appearing on the scene.

Evolutionists are so focused on promoting their delusional atheist agenda of convincing the public there wasn't a Creator they fail to see the obvious. Proven facts show the extinction rate of species world-wide is escalating and the world is degenerating. It's not getting better from random mutations; new species are not appearing; they are rapidly dying off. And that's not a theory, here is evidence.

The Center for Biological Diversity claims:

"In the past 500 years, we know of approximately 1,000 species that have gone extinct, from the woodland bison of West Virginia and Arizona's Merriam's elk to the Rocky Mountain grasshopper, passenger pigeon and Puerto Rico's Culebra parrot — but this doesn't account for thousands of species that disappeared before scientists had a chance to describe them. Nobody really knows how many species are in danger of becoming extinct. Noted conservation scientist David Wilcove

*estimates that there are 14,000 to 35,000 endangered species in the United States, which is 7 to 18 percent of U.S. flora and fauna. The IUCN has assessed roughly 3 percent of described species and **identified 16,928 species worldwide as being threatened with extinction**, or roughly 38 percent of those assessed. In its latest four-year endangered species assessment, the IUCN reports that the world won't meet a goal of reversing the extinction trend toward species depletion by 2010."*

If evolution is true, why is the extinction rate so high? They tell us species evolve for survival, yet statistics and history tell us that's simply not true; it's actually the opposite, species are dying off and no new varieties are appearing, and that is fact.

Evolution's high priest Richard Dawkins, who can bore you for endless hours with his rehearsed repetitive rhetoric on all his theories of evolution, will probably use the standard evolutionist escape phrase used so many times it no longer has any credibility. It goes something like this... *'evolution happens over millions of years so even if the bees became extinct some other specie will appear on the scene to save our food supply so all will be ok, just believe in evolution, it will solve all of humanities problems'*.

Pity help us if evolutionists ran the world. Whilst they wait millions of years for evolution to do its miraculous all-knowing evolutionary work, mankind will run out of food and we will all starve. Mankind doesn't have millions of years to wait for a replacement for bees if food sources run out.

Evolutionists also have another theory that I refer to as their 'doomsday theory'. Evolution's plan is for humanity's extermination, so another version can evolve from another fish or whatever is superior and doesn't need food to survive. Yes, that's another one of their theories to make you feel better.

Anything is possible in the minds of evolutionists. They invent endless theories, regardless of how crazy they sound. If it sounds good enough and their peers also like the idea, then it becomes an established fact in their delusional science fiction minds.

Evolutionists also tell us all the species we have today evolved from other life forms. If true, why don't we see new species appearing. Why is the total number of species on earth in rapid decline? There are no new species appearing or evolving because it doesn't happen that way, the undeniable facts prove this.

Where is the evidence from evolutionists on newly evolved species? Once again, they don't have an intelligent reply other than their favorite copout answer, *'it takes millions of years'*. How can scientists be so naïve?

Evidence shows that based on the current species extinction rate, in a million years none will survive, including us as there will be nothing to eat.

Here is another simple example why evolution is impossible. They claim many animals evolved from fish by evolving into amphibians and sprouting legs, growing lungs etc. Somehow, they changed their bodies to survive on land.

The Center for Biological Diversity states:

*"**No group of animals has a higher rate of endangerment than amphibians**. Scientists estimate that a third or more of all the roughly 6,300 known species of amphibians are at risk of extinction. The current **amphibian extinction rate may range from 25,039 to 45,474 times the background extinction rate.**"*

"Frogs, toads, and salamanders are disappearing because of habitat loss,

water and air pollution, climate change, ultraviolet light exposure, introduced exotic species, and disease. Because of their sensitivity to environmental changes, vanishing amphibians should be viewed as the canary in the global coal mine, signaling subtle yet radical ecosystem changes that could ultimately claim many other species, including humans."

Based on these numbers, amphibians could not have survived billions of years. With their high sensitivity to environmental changes it would be impossible for them to have survived the toxic environment evolutionists claim existed on the early earth. They would have all become extinct before evolving into amphibians. Their survival requires a perfectly balanced earth and environment, a scientific fact.

With evidence proving amphibians would not survive in an early earth, this creates another monumental problem for evolutionists who claim we evolved from amphibians.   What happens when you take a fish out of water or remove a frog from a water supply, they die in a very short time. Now add in a toxic environment and sensitive amphibians could never survive.

Can you really believe you came from amphibians? Then, just by a natural progression you eventually developed a human body and superior intellect to animals. The human body is so complex, scientists don't know a fraction of how the brain really works, yet somehow this magical force of evolution, with no intellectual form, designed the human brain and supervised its development. DNA, the code of our body is so incredibly complex that believing we developed by random chance defies basic common-sense reasoning. In chapter six you will discover some incredible facts on DNA, also scientifically proving evolution is impossible.

Where is the evolutionists' proof, beyond their meaningless unproven theories of new species arriving on the scene? Where is the proof of frogs and toads changing into a new species to survive? There are none. Fact is, species are rapidly dying, becoming extinct and earth is degenerating.

Here are more facts.

Birds

A 2009 report on the state of birds in the United States found that 251 (31 percent) of the 800 species in the country are of conservation concern]. Globally, BirdLife International estimates that **12 percent of the known 9,865 bird species are now considered threatened, with 192 species, or 2 percent, facing an "extremely high risk" of extinction in the wild.**

Fish

The American Fisheries Society identified 700 species of freshwater or anadromous fish in North America as being imperiled, amounting to 39 percent of all such fish on the continent. In North American marine waters, at least 82 fish species are imperiled. **Across the globe, 1,851 species of fish — 21 percent of all fish species evaluated — were deemed at risk of extinction by the IUCN in 2010, including more than a third of sharks and rays.**

Invertebrate

Of the 1.3 million known invertebrate species, the IUCN has evaluated about 9,526 species, **with about 30 percent of the species evaluated at risk of extinction.**

Mammals

*The IUCN estimates that half the globe's 5,491 known **mammals are declining in population and a fifth are clearly at risk of disappearing forever with no less than 1,131 mammals across the globe classified as endangered, threatened, or vulnerable.** In addition to primates, marine mammals — including several species of whales, dolphins, and porpoises — are among those mammals slipping most quickly toward extinction*

Plants

*Of the more than 300,000 known species of plants, the IUCN **has evaluated only 12,914 species, finding that about 68 percent of evaluated plant species are threatened with extinction.***

Reptiles

*Globally, **21 percent of the total evaluated reptiles in the world are deemed endangered or vulnerable to extinction** by the IUCN — 594 species — while in the United States, 32 reptile species are at risk, about 9 percent of the total. **Island reptile species have been dealt the hardest blow, with at least 28 island reptiles having died out since 1600.***

Hard cold evidence species are rapidly becoming extinct, they don't and can't evolve to save themselves, further proof evolutionists are conning you.

Darwin recognized this, and to the dismay of evolutionists, he admits species can't change fast enough to survive and are exterminated. Even he acknowledges the theory's many flaws.

Though nature grants vast periods of time for the work of natural selection, she does not grant an indefinite period; for as all organic beings are striving, it may be said, to seize on each place in the economy of nature, if any one species does not become modified and improved in a corresponding degree with its competitors, it will soon be exterminated. . (Darwin, The Origin of Species.)

The Genius of Darwin, as evolutionists claim, was not evident in his comments;

"Our ignorance of the laws of variation is profound. Not in one case out of a hundred can we pretend to assign any reason why this or that part differs, more or less, from the same part in the parents. ... why do we not find them embedded in countless numbers in the crust of the earth?" (Darwin, The Origin of Species.)

Out of the thousands of specimens Darwin collected, he couldn't find a single one that proved transmutation, hence his above comments.

Evolutionists, much to their annoyance can't find a single verified sample and have a gaping hole in their theory.

All they can do is invent theoretical drawings of evolutionary stages in a specie concocted in their minds; but they are not true or evidence, they are pure science fiction.

Darwin desperately tried to come up with evidence. He pondered on the Mustela (American Mink) of North America.

"Look at the Mustela vison of North America, which has webbed feet and which resembles an otter in its fur, short legs, and form of tail; during summer this animal dives for and preys on fish, but during the long winter it leaves the frozen waters, and preys like other polecats on mice and land animals."

This is no evidence of a transmutation specie, a bear preys on fish in summer, they gorge themselves on salmon to fatten up for hibernation.

A creationist would state God made the American Mink as it is, a perfect design for its environment, and the evolutionist could not prove them wrong. They have no proof of its transmutation.

There are animal species that evolutionists could claim are transmutational such as a flying squirrel. Is it a bird or a rodent, they can't prove it's in a transmutational state and the creationist states it's one of God's perfect designs which evolutionists can't disprove. I could go on with many other examples, however I'm sure you get the picture. They cannot prove any species living today is transmutational.

Darwin lists many other species he observed and hypothesizes how they may have evolved, but he never proved any evidence to back up his claims. He assumed, he guessed, he theorized, he dreamed, he wished, he wanted... despite there being no evidence. Evolutionists twist Darwin's words and findings.

Evolutionists never mention Darwin's struggles with the evolution of complex items such as the human eye, or the eagle's eye. Darwin wrote he couldn't believe it could form through natural selection. Even he knew that was too far-fetched to be possible, so he was forced to acknowledge the work of a Creator.

"To suppose that the eye, with all its inimitable contrivances for adjusting the focus to different distances, for admitting different amounts of light, and for the correction of spherical and chromatic aberration, could have been formed by natural selection, seems, I freely confess, absurd in the highest possible degree."

"...that a living optical instrument might thus be formed as superior to one of glass, **as the works of the Creator** *are to those of man?"*

(Charles Darwin, The Origin of Species, chapter 6)

Evolutionists never refer to that confession from Darwin, as it kills evolution.

In the 1800s Darwin knew nothing about other complex structures of the human body. He had no idea how complex the human eye is and the impossibility of it evolving. Modern medical science has also proven evolution impossible.

Biochemist, Professor Michael Behe details in his book, *Darwin's Black Box*, how vision works when light hits the eye and is converted to vision. It's incredible. The detail and complexity make it clear as day. The eye could never have evolved from random mutations and mere chance. It's just basic common-sense to see the impossibility of this scenario. He summarizes by stating it is impossible for vision to have evolved because all the intricacies of the chemistry of the eye must be present at the same time for it to work. He states scientifically it's impossible for bits to appear first and then get better through evolution. Vision can only work if all the components are present at the same time.

It's like turning on a light switch. For it to work, the wiring must be connected, the globe has to be perfect with all the components in place and functioning. If one part isn't in place, the light globe won't work. The same is true with the eye, everything had to be in place from day one. It was impossible for it to evolve; it had to be created in its entirety for it to work.

Behe also describes how other parts of the body function, clearly showing the various body systems had to be in place in their entirety at a specific point of creation for the human body to function. It had to be created. This is real science proving evolution isn't true and couldn't work, these are living scientific proven facts, not theory.

Darwin, who evolutionists revere as the discoverer of evolution and the proof there is no God, acknowledges the work of the Creator. You will never hear atheist evolution scientists mentioning this fact. Darwin struggled with the complexity of human design, the variety and complexity in the animal kingdom, and the impossibility of evolution being the force at work creating such complex and working designs.

Darwin had his thoughts which he put on paper, that's the bulk of his writings, they were unproven ideas. Darwin even admitted he found it hard to believe the varieties of pet pigeons he kept, and after studying them for years, had difficulty believing they could have descended from a common parent. These are the words from the god of the evolutionists upon whom they credit with discovering evolution.

Below is yet another of Darwin's admissions that he actually discovered his own theory of evolution impossible.

*"I have discussed the probable origin of domestic pigeons at some, yet quite insufficient, length; because when I first kept pigeons and watched the several kinds, knowing well how true they bred, **I felt fully as much difficulty in believing that they could have descended from a common parent, as any naturalist could in coming to a similar conclusion in regard to the many species of finches, or other large groups of birds, in nature.** One circumstance has struck me much; namely, that all the breeders of the various domestic animals and the cultivators of plants, with whom I have ever conversed, or whose treatises*

I have read, are firmly convinced that the several breeds to which each has attended, are descended from so many aboriginally distinct species. Ask, as I have asked, a celebrated raiser of Hereford cattle, whether his cattle might not have descended from long-horns, and he will laugh you to scorn. I have never met a pigeon, or poultry, or duck, or rabbit fancier, who was not fully convinced that each main breed was descended from a distinct species. Van Mons, in his treatise on pears and apples, shows how utterly he disbelieves that the several sorts, for instance a Ribston-pippin or Codlin-apple, could ever have proceeded from the seeds of the same tree. Innumerable other examples could be given."

Darwin calls his book 'Origin of Species', *"one long argument"* and that's exactly what it is. It's not the discovery of evolution and it's certainly not the work of a genius as the evolutionists claim. In the 1800s Darwin's level of knowledge was primitive compared to today's medical science.

"The noble science of Geology loses glory from the extreme imperfection of the record. The crust of the earth with its embedded remains must not be looked at as a well-filled museum, but as a poor collection made at hazard and at rare intervals. The accumulation of each great fossiliferous formation will be recognised as having depended on an unusual concurrence of circumstances, and the blank intervals between the successive stages as having been of vast duration?" (Darwin, The Origin of Species.)

Darwin died from a massive heart attack in 1882. He was born into a wealthy family, never having to work he had too much spare time to indulge in fantasizing on how species formed. In the decades following his death, Darwin's thoughts have been twisted and distorted by atheist evolution scientists using his writings as their bible.

Evolutionists frequently misquote him, claiming his theories are science and proven. In reality, Darwin proved nothing.

Whilst Darwin had his group of like-minded cronies, mainly in allied academic fields, he also had many critics, which evolutionists seldom mention as it would tarnish Darwin's credibility as their god.

The evil legacy of Darwin's writings

While most discussions on Darwin focus on evolution, there is also a dark side to Darwin's books. Applying Darwin's theories to humanity intrigued his half cousin Francis Galton, who suggested humans could be bred to encourage positive physical and mental traits similar to what evolution espouses in nature.

Darwin supported the idea in theory, but not in practice because he believed it would stymy the best in humanity. Galton would later take these ideas and create 'eugenics', the science of improving a population by using controlled breeding to increase the occurrence of desirable heritable characteristics. This is what Hitler tried to do by killing 6 million Jews and 5 million, Poles, Roma (gypsies), Soviets, homosexuals, blacks, physically and mentally ill that Hitler racially viewed as biologically inferior.

Hitler believed the survival of the German "Aryan" race depended on its ability to maintain the purity of its gene pool. It is clearly documented that Hitler had many scientists and philosophers working on his holocaust experiments to evolve the perfect race. Evidence is available outlining the horrific medical experiments Hitler's scientists undertook to further the cause of eugenics.

Another leading advocate for eugenics in America was Margaret Sanger, the founder of Planned Parenthood, the world's largest abortion organization. Sanger's thoughts on the inferiority of certain elements of society, especially the black race is well-documented. Her calls for birth control and abortion were to weed out "undesirables" such as those with birth or mental defects. To this day, Planned Parenthood's highest honor is called the Margaret Sanger Award and is awarded annually.

Evolutionists desperately try to separate themselves from the fact that Darwin's books influenced Hitler. However, they cannot get away from the fact that writings from leading Nazis and German biologists often referred to Darwin's theories. The Nazis followed his theory on survival of the fittest and elimination of weaker specie through natural selection, under which they instigated the wholesale murder of millions in gas chambers and other barbaric ways. Darwin taught that weaker animals get killed by the strongest and only the strongest species survive. There is extensive documentation that Darwinism influenced Nazi policy.

In *Origin of the Species*, Darwin wrote:

"As natural selection acts solely by the preservation of profitable modifications, each new form will tend in a fully stocked country to take the place of, and finally to exterminate, its own less improved parent or other less-favoured forms with which it comes into competition. Thus, extinction and natural selection will, as we have seen, go hand in hand. Hence, if we look at each species as descended from some other unknown form, both the parent and all the transitional varieties will generally have been exterminated by the very process of formation and perfection of the new form." (Darwin, The Origin of Species.)

In Hitler's book, *Mein Kampf*, he mentioned Darwinism was a fundamental basis for achieving a successful German superior race of humans, and he believed selective breeding could improve German's genes. This is common knowledge and fact is the Nazis relied heavily on Darwin's writings for their design of the 'Superior Race'. There are many quotes from Hitler on the subject, it's recorded in history but conveniently evolutionists deny this inconvenient truth or claim he "misunderstood" what Darwin meant.

Hitler began reading about eugenics and social Darwinism while imprisoned, following a failed 1924 coup attempt known as the Beer Hall Putsch.

The following is from Darwin's book Descent of Man, supporting eugenics and his views on the elimination of perceived weaker species including humans. Curiously, evolutionists never mention this book. This was also Hitler's views which he may have gotten from studying Darwin's books.

"We civilised men, on the other hand, do our utmost to check the process of elimination; we build asylums for the imbecile, the maimed, and the sick; we institute poor-laws; and our medical men exert their utmost skill to save the life of every one to the last moment. There is reason to believe that vaccination has preserved thousands, who from a weak constitution would formerly have succumbed to small-pox. Thus the weak members of civilised societies propagate their kind. No one who has attended to the breeding of domestic animals will doubt that this must be highly injurious to the race of man. It is surprising how soon a want of care, or care wrongly directed, leads to the degeneration of a domestic race; but excepting in the case of man himself, hardly anyone is so ignorant as to allow his worst animals to breed." (Darwin, Descent of Man)

Hitler adopted social Darwinism, the survival of the fittest.

He believed the German master race had grown weak because of the influence of non-Aryans (people of Indo-European heritage) in Germany.

His Masterplan, *'Generalplan Ost'* was to exterminate or enslave the Slavic population, then conquer and repopulate their lands with his superior German settlers.

History also records many other of the world's most evil men studied Darwinism. Stalin, Karl Marx, Lenin, and Mao Tse-tung, who probably holds the title of the most-evil human, all embraced Darwinism. The number of deaths under his tyrannical reign is staggering. There are numerous books on these mass murderers and their link to Darwinism which you can read if you have further interest in this subject.

Naturally, no one holds Darwin directly responsible for the actions of those who came after him. At the time of his writings he had no idea who would end up reading his books and the effect his theories would have on the world.

It defies common-sense and moral responsibility for atheist evolution scientists to elevate Darwin as the god of evolution. It is very concerning they claim how wonderful he was, using absurd words like 'a brilliant mind', when his writings became the seed for much harm and suffering, including the largest death toll in the history of mankind during the 20th century. It is morally corrupt for evolutionists to think highly of Darwinism.

One would like to think if Darwin had known the Nazis would use his theories to justify the killing of 11 million innocent humans just because they were of a different race, he would never have written his books, but one cannot say this with certainty.

Darwin was born into a privileged English environment, supported by his wealthy father and inheriting significant wealth. He lived in a privileged, sheltered environment where he never had to worry about earning money. As part of this aristocratic upbringing, Darwin thought himself superior, as evidenced by racial comments that, if made today, would get him permanently barred from society, yet evolutionists still revere him as their god.

"If a naturalist who had never seen such beings, were to compare a Negro, Hottentot, Australian (aboriginal), *or Mongolian, he would at once perceive that they differed in a multitude of characters, some of slight and some of considerable importance. On inquiry he would find that they were adapted to live under widely different climates, and that they differed somewhat in bodily constitution and mental disposition."*

"...and the increased size of the brain from greater intellectual activity, have together produced a considerable effect on their general appearance in comparison with savages.

"At some future period, not very distant as measured by centuries, the civilised races of man will almost certainly exterminate and replace throughout the world the savage races."

(Charles Darwin. The Descent of Man and Selection in Relation to Sex, Vol. I)

Sorry ladies, he also made highly offensive and sexist comments about you.

Darwin claimed men are superior to women. Women are inferior, at a lower level of development than men with smaller brains, eternally primitive, childlike and a danger to contemporary civilization.

Imagine if someone made those statements today. They would rightfully be condemned at all levels, their books banned, and they would be ostracized by society. Yet evolutionists claim Darwin was a brilliant mind and they revere him as their god. Darwinism is taught in schools; the teaching of this racist, misogynistic man should be banned from every educational institution.

Darwin's poor health prevented him from socializing or debating his theories with his peers, forcing him to remain at home. This is an important factor in understanding his writings. He wasn't actively researching specie. When he wrote about species in other countries, it was all textbook stuff. Apart from his Beagle journey, Darwin never travelled overseas. Many of the specie he wrote about weren't from personal observations. Instead, he read about them and made assumptions based on other people's observations, which further questions the originality of a lot of his work.

Darwin was also a prolific reader of other naturalists like Jean-Baptiste Lamarck (1744–1829), a French naturalist who proposed biological evolution in 1801, decades before Darwin. Lamarck theorized that all organisms become more complicated over time, and therefore it doesn't account for simple organisms such as single-cell organisms.

Lamarck believed a giraffe's neck grew longer during its lifetime as it stretched to reach leaves in high-up trees after eating all the low-hanging leaves. Meaning each generation of giraffe had a longer neck than previous generations. That didn't work out for him.

Lamarck's theory *of Inheritance of Acquired Characteristics* was proven wrong through genetics, as science develops it proves theories wrong. As we know, a theory is simply an unproven idea.

External changes made to an animal cannot be passed to their offspring. If a vet operates on an animal, changing its ears or tail, the parent doesn't pass the changes to its offspring.

Through observations of species, Darwin claimed parents passed traits to offspring, but he never understood how they were passed on as genetics wasn't discovered in his era.

Today, we now know a lot about genetics and the only way we pass traits to the next generation is through the genetic code which cannot be affected by natural forces such as the mysterious force of evolution.

Medial science has discovered gene editing and we can now disable target genes to correct harmful mutations causing illnesses and human diseases. This has nothing to do with evolution. It's a positive scientific discovery helping mankind, this is where science can be so productive. The danger with gene modification would be if they go too far. We read articles alluding to gene modification for designer babies, this is very dangerous as we don't know what further mutations could affect subsequent generations.

Gene coding poses a major problem for evolutionists. If parents pass on traits to their offspring, how are they changed to enable the species to evolve since outside natural environmental factors or elements can't change genes.

The only thing that can change offspring is the gene sets passed on from the parents.

I referred to this in the previous chapter. Do you know how to change the gene set in your future children? Can you mentally influence the gene code in your children so they will evolve into super intelligent individuals, a top golfer, or whatever else you desire? I'm guessing your answer is the same as mine, no. If, miraculously you do, bottle it. You will sell more than Coca-Cola and make billions.

We know alterations in an offspring's gene code cannot be willfully instigated by the parents, this is a MAJOR problem for evolutionists. How is the evolution plan for a specie passed from one generation to another?

They will probably use their lame fallback answer again, *'it happens over billions of years, just ever so slightly through each generation so you can't see it, science has proved it'*, which would be a lie, they haven't and can't prove it.

Evolutionists know the public has been conned for so long into believing their crazy statements they can now say whatever they like with impunity as they are never asked to prove it.

Evolutionary theory teaches human intelligence has evolved. Weaker inferior humans die off while the strong survive through some unexplainable natural evolution of the gene code, this is the basis of evolution.

Are humans more intelligent today? We can't go back a million years as there are no records of humans. We can however go back to remaining testable examples of their achievements, real stuff, not theories.

Let's consider the Pyramids of Egypt. Even with today's technologies it's amazing how 4,500 years ago the Pyramids were constructed using only manual labor. Even by today's standards they are an engineering masterpiece.

The level of intelligence used to create the pyramids without modern equipment is impressive. It could be claimed their designers and builders were more intelligent than humans today with their help from computers and modern technology.

You be the judge, were the pyramid builders highly intelligent?

- The pyramids were the tallest man-made structures on earth for 3,800 years.
- Today there are a reported 5,000 pyramids still standing.
- Every stone weighs 5-10 tones and was manually lifted into

place, no cranes or forklifts. Researchers are still trying to figure out how they did it as pyramids could be 480 feet high, the height of a 50 story building.

- Internally they built an intricate maze of tunnels, passages, air shafts and chambers.

- Despite searing desert day temperate and cold nights, the pyramids maintained a constant 20-degree Celsius inside, the best air-conditioning, didn't need electricity and cost nothing to operate.

- They aligned tombs north to south with an accuracy of up to 0.05 degrees. This happened in an era where modern-day precision tools were not even imaginable!

- One reason the pyramids have stood for so long is the super adhesive mortar used to fix the stones in place. Analysing them still haven't revealed the combination of the ingredients, it remains a mystery. Almost 5,000,000 tons of mortar were used, and their strength is more than the stones themselves.

- There are 3 huge swivel doors in the pyramid of Giza. One weighs 20-tonnes and is so well balanced it can be opened from the inside with just one hand! When closed it's invisible as it blends in perfectly with the wall.

- In the Great Pyramid of Giza, the descending shaft or passage is aligned exactly to the polar star 'Alpha Draconis' when built in 2141 BC. Over 4100 years ago.

- The polar star comes in alignment with earth once every 25,920 years. When the pyramid was being built, the architect used the alignment of Alpha Draconis to determine the direction to true north.

- At the time of construction, they covered the exterior with a well-polished casing of limestone which sparkled like diamonds when the sun was shining, making them visible from miles away.

- The three Giza pyramids align exactly with the three stars in the Orion belt, pointing directly towards Orion's belt.

- The Great Pyramid of Khufu weighs a staggering 5,750,000 tons! As a comparison, the Burj Khalifa, the world's tallest building at the time of this writing, weighs only 500,000 tons!

- The 'Great Pyramid' was built from an estimated 2.3 million blocks of stone, and each block weighs 2267.96 kilograms (2.5 tons) approximately. Do the math!

- Though highly sophisticated in terms of accuracy and architecture, the Great Pyramid of Giza took almost 23 years to complete. Most of the pyramids were constructed simultaneously.

- In 1877, Dr. Joseph Seiss made an interesting discovery. He showed that they located the pyramid of Giza exactly at the intersection of the longest line of longitude and longest line of latitude, which means it is in the exact centre of all landmasses on planet Earth. Coincidence? Or were these highly intelligent human beings? How did they work that out with no computers, Google Earth or satellite imagery?

- The pyramids are one of the marvellous examples of engineering from ancient history and it is the only surviving structure of the Seven Wonders of the Ancient World.

The pyramid builders were extremely intelligent and I'm sure most builders and engineers today would envy their level of intellect. Builders and engineers would probably admit that if they had to use the same tools available 5000 years ago they couldn't build these structures. There is an argument to be made that perhaps humanity isn't getting smarter, especially when you listen to evolution scientists who claim to be intelligent then try to tell us we all came from a fish. You be the judge; I choose the Egyptians in that comparison.

Does common-sense reasoning tell you something seems very fishy with the evolution scientist's fishy tale, how one day by random chance a fish of some type decided it wanted to become a land mammal. What type of fish? They don't know exactly, if it was millions of years ago it could have been a jellyfish, but the problem with that is they don't have a brain?

One scientist came up with what he calls a 'sea creature called a Tiktaalik'. After finding some fossilized remains in the arctic he came up with yet another theory that this is the fish we evolved from billions of years ago, once again, no proof.

Whenever theses vague unproven discoveries are made, evolution scientists go into overdrive, making outlandish claims such as, 'we now know how we evolved from a fish, we have discovered the

species of fish we all came from because it has fossilized structures that will ultimately become parts of our human bodies, shoulders, elbows, legs, a neck, a wrist — they're all there in the Tiktaalik.'

In fact, they took it even further, claiming:

It's like peeling an onion," he says. "Layer after layer after layer is revealed to you. Like in a human body, the first layer is our primate history, the second layer is our mammal history, and on and on and on and on, until you get to the fundamental molecular and cellular machinery that makes our bodies and keeps are cells alive, and so forth.

This is rubbish, how could they possibly discover all that from fragments of a fossil they claim is 550 million years old, another unproven claim. Then, how can they peel off layers like an onion to reveal our primate history. He claims this piece of fossil shows all the stages of humankind's evolution. Really, it's a fossil with all those layers. Dating is highly suspect if in fact it was dated. Who knows? They make so many untrue claims. You will read in a subsequent chapter how fossil dating is also flawed. The claims these guys make are just so ridiculous. They know they can say anything to achieve notoriety with their like-minded cronies because no one ever asks them to prove it.

The Tiktaalik inventor even sketched it, claiming it was an accurate representation of what the sea creature MAY HAVE looked like. Yes, it's a 'may have'. I wish they would all stick to the same script. There are so many theories and opinions from these guys just on the fish we evolved from, this in itself shows no credibility.

A fish brain is one-fifteenth the mass of a similarly sized bird or mammal. Compared to the human brain it is very undeveloped and infinitely smaller, it's not even remotely similar. It's like saying an ancient abacus changed itself into the most sophisticated supercomputer known to mankind all by itself.

Somehow the fish and all its subsequent offspring over the next 550 million years changed their genes in an organized and orderly way. Remember, genes are very intricate and every piece has to be in perfect sequence. If one bit is in a wrong order it can break down its immune system, so it becomes extinct from bacterial attack.

For these innumerable changes over millions of years to happen in perfect order with no intelligent being designing and controlling the process and nothing going wrong is cuckoo. The mind boggles at the impossibility of all this happening. Is common-sense reasoning telling you how crazy this stuff is?

Its science fiction and this is what the evolutionist scientists continually spout forth, making rash claims that it's all true when it's unprovable and only a theory.

The final sentence goes to Darwin confirming he believed in creation.

*"...yet every naturalist knows vaguely what he means when he speaks of a species. Generally, the term includes the **unknown element of a distinct act of creation.**"*

The Big Bang

What is The Big Bang? Who discovered it? Is it provable by using the scientific method? Is it true?

Most people have heard of 'The Big Bang', but few understand its origin and the science behind it. Scientists have incorrectly referred to it as proven science so often that many people now believe it's a proven scientific fact.

Evolution scientists tell us everything started with 'The Big Bang'. Once upon a time (all good fairy tales start with that line), around 13.7 billion years ago there was a monumental explosion. This catastrophic event set off a process that led to an infinite

number of random events and mutations, thus creating everything; the universe, including the moon, sun, stars, earth, animals, and even you and me. The fairy tale gets better. With no intelligent designer, everything miraculously fell into place in perfect order, nothing went wrong and from this chaos a perfect universe appeared.

They claim that after this initial explosion gases were left behind. They don't know what these gases were as there are no remaining samples, so they just make an educated guess what they may have been, conveniently only selecting those gases that fit their theory. These assumed gases somehow managed to mix together in just the right quantity and order and then poooooof. Just like a magic show we ended up with all the various elements we have today, including oxygen, otherwise you wouldn't be reading this book.

Obviously, no one witnessed this event so it cannot be scientifically proven. There is no proof, it's only a theory invented by evolution scientists who continually create science fiction theories full of so many holes they could fill all the black holes in outer space!

For scientists to call the Big Bang a theory is incorrect and misleading to begin with, it's only a hypothesis.

The dictionary definition of theory is:

a supposition or a system of ideas intended to explain something, especially one based on general principles independent of the thing to be explained.

The dictionary definition of hypothesis is:

a supposition or proposed explanation made on the basis of limited evidence as a starting point for further investigation.

The Big Bang is only a hypothesis as there is NO EVIDENCE and scientists cannot prove it happened using the scientific method or recreate it, hence it's only a hypothesis.

Evolutionists even use an incorrect term for it intentionally to mislead and con the public.

First, let's run the premise of the Big Bang past our common-sense credibility test.

Before this initial big explosion, goodness knows how the explosion happened or where these mysterious gases to fuel it came from, according to evolutionists there was nothing before this event. That's what some of them claim, but to confuse us even further there are a variety of theories about the specifics and they all claim to be right, so who really knows!

We know something cannot come from nothing, that's common-sense logical reasoning, so such an explosion is impossible. So much for that theory. If there was an explosion, something must have existed before it happened. If so, no one actually knows who or what that was. If anyone claims they do, they are not being honest for it's impossible to know. It would just be another of their unproven theories.

However, keeping to their fairy tale theory; after this big explosion, leftover gases that came from nothing and just happened to be the right ones, and in the right quantities were floating around at just the right time. Then, in a super-miraculous way, these gases somehow joined and magically formed the very first cell.

I'm not a mathematician, but the math behind the odds of all those factors coming together in perfect order would be bigger than the biggest number known to mankind, which is a 'Googolplex'.

A Googol is 10^{100} (that's a 1 followed by 100 zeros), and a Googolplex is 10 to the power of a Googol. The number would be greater than any number ever used, which means we can create a new number. If it catches on and used, we have created the biggest number, that's how they invent big numbers. So, let's create a new number and call it a 'Googolbang', meaning it has so many zeros that if you loaded it into a computer it would probably blow up, just like the Big Bang. This would be a very fitting number.

They claim the earth, sun, moon, stars, and everything in the universe formed when all the clouds created from the Big Bang began compressing. Yes, that's the theory, no proof, just a theory. More on that impossibility later.

From this joining of gases came life forms, referred to as 'chemical evolution'. Everything originated from this random marriage of a bunch of anonymous gases meeting in space one day and saying to each other, let's join together and make cool stuff. We can get married and form a living cell, then one day our offspring can form earth and all living life. That's their fairy tale.

If I offered a compulsive gambler a hundred billion-to-one odds of that happening, they would have more sense to pass on those odds and spend their $1 at McDonalds.

Ponder the odds. Gases organized themselves into the perfect composition, types, and amounts at exactly the right time. This environment had to be the perfect temperature for this reaction. Remember, this didn't happen in a controlled laboratory. A force had to mix the gases and somehow hold them together to react in the vast void they claim existed. Millions of other variables also had to perfectly fall into place.

All this supposedly happened without a brain working out the formula, directing or controlling the mixing. These combinations and components all miraculously came together perfectly, mixing at precisely the right time and in the right place.

Somehow, something with no life turned into a living cell able to look after itself in this highly toxic environment scientists tell us existed back then. Living cells need a food source and a perfect atmospheric environment to survive.

This original cell went on to somehow reproduce itself. Goodness knows how as there was just one cell. How did it copy itself? How was a DNA code formed so it could copy itself and multiply? In a following chapter, you will learn how this is impossible.

Then, as this family of simple cells expanded, they continued evolving into ever more complex cells; multi cells which worked together like a well-organized army. No one knows how their DNA code changed, but we won't dwell on that. Perhaps they had highly superior cell brains millions of times more sophisticated than humans so they could develop and change their own DNA and evolve into a new crawling thing that never existed before. It was just like magic. It's important to note that when they invented the Big Bang fairy tale they didn't know about the complexity of DNA, which makes their cell mutation theory impossible.

The fairy tale continues. Over millions of years each cells' offspring continued the evolutionary process; making changes so minute you don't notice them from one generation to another. That's the evolution theory. Miraculously, all these minute changes continue building up as each generation somehow passes the required pattern for change to the next generation with no flaws. This is all happens with no intelligent designing force directing it.

The entire evolution process hinges on all these millions of generations of random mutations randomly coming together in perfect order. It had to be perfect. If one error entered the mutation process, the poor little cells would become extinct. If a single DNA code change is wrong, they die. Evolutionists teach that if a species develops a weakness in its DNA, the force of evolution ultimately culls it out, causing this species to become extinct. This is a major problem for the very early days of their theory of evolution as they claim there were no mistakes. That's impossible. Their theory self-destructs at this early stage.

However, let's assume (which is the evolutionist's favorite word) it all goes according to their theory and these little cells grow up and sprout all the different attributes needed to survive. Oh, that's right; for all this to happen they need planet earth.

Let's not forget all the stars and planets. Especially how our sun and moon have been hung in space at just the right location and functioning at just the right time. If the sun was slightly closer to earth, those poor little cells would be grilled. Any further away and earth would be a big ice block. If the other planets weren't fixed in the perfect orbital path, one of the millions of other stars out there would have destroyed earth a long time ago.

This brings up other BIG questions. How did the planets and stars just happen to become fixed in exactly the right places for the universe to work? How do they follow their invisible pathway without crashing into each other? How did the orbit of planets evolve?

If anyone believes all this happened by random chance from a big explosion, it's time for a serious reality check.

Ponder on all the millions of crawling things which supposedly came from simple lumps of jelly and, to survive, had to rapidly evolve into lifeforms able to protect themselves in the super harsh toxic environment of early earth. It's science fiction.

The fairy tale continues. These lumps of jelly decided it was boring being a lump of jelly, so without knowing what a fish was, as fish didn't exist, they planned to evolve into a fish. Yes, a crawling thing with the brain the size of a pinhead designed and built a fish.

Even humans with the highest level of intelligence of any specie and harnessing the power of supercomputers and artificial intelligence cannot invent a new species as complex as a living fish, or animals capable of surviving on their own. They can't create life from nothing, yet evolution scientists tell us that lumps of jelly designed themselves into a fish.

Then the fish got bored just swimming around in water and wanted to be a mammal. While living underwater they designed body parts never previously seen, lungs, heart, reproductive systems, and changed fins into arms and legs so they could move onto the land. They continued their superior design and building talents, eventually waking up one day as a huge dinosaur. Give me a break. This is better fantasy reading than Hollywood can ever produce.

What's the evidence? Just drawings of the assumed stages and a few random highly suspect fossils. They have no concrete evidence of these species transitioning through the various stages linking events together. All they are able to provide are drawings based on their assumptions of how it may have happened. To the evolutionist that isn't a problem as they treat any theory invented by any scientist as being true.

Evolution scientists know they can say anything as they are never asked to prove it with empirical evidence.

The word 'theory' means 'assumptions', not fact. Do you know what the word 'assumption' breaks down to, it means to make an *'ass -out-of-u & me'*, which evolution scientists continually do with their endless theories.

They tell us life formed from nothing, no intellectual force, no designer, no planner, no creator, this marvelous thing happened all on its own just by chance.

And all this started from a monumental explosion. My experience with explosions is they destroy, they don't create.

As a kid, we enjoyed an annual event called "Guy Fawkes. This English celebration involved Guy Fawkes, real name Guido Fawkes. In the 1600s Fawkes, along with 13 co-conspirators, planned to blow up King James and the English parliament during the opening ceremony. They tunneled under parliament, where they placed 40 barrels of gunpowder to create another big bang. Unfortunately for them, their plot was uncovered.

Poor old Guy was caught red-handed, matches in hand. Needless to say, things didn't end too well for him. King James and parliament survived; however, Guy didn't see another opening of parliament. The King had him hanged. Afterwards, they encouraged their countrymen to celebrate the King's survival, so since then, on each anniversary of the opening of parliament they light bonfires to celebrate. It's like a Thanksgiving Day for the English. As part of the festivities they placed an effigy of Guy on top of the bonfire. Fireworks were added a few years later.

Kids would make an effigy of 'Guy' from flammable material such as paper or straw. They would then place him in a cart and walk around the neighborhood yelling, "guy, guy, guy, penny for the guy". The money raised was used to purchase fireworks for bonfire night or your own backyard adventures.

Leading up to the big night, the neighborhood built a huge bonfire on the local beach using anything flammable they wanted to dispose of. This is a great time of the year to prune trees, remove old fences, and clean out the basement.

On Guy Fawkes night the neighborhood gathered around and hoisted the guys to the top of the bonfire, while one lucky person was privileged to throw in the match. With the help of some gas, poooooof, bonfire heaven. And yes, we always got a permit from the fire department. Some people today may frown at the event, but as kids we had a great time, far better than Halloween. In all the years we did it, I can't remember anyone being injured. An odd burn on the hand didn't count. I think there was more common sense back then and any sick firebug got it out of their system.

I thought you would enjoy this light-hearted bit of British history and there was a reason for it.

Every year, on sale was a huge range of fireworks, not just pretty ones. We had 'crackers' that blew up, some had names like 'Mighty Cannon' and went off with a booooom, blowing its casing into a million pieces. I hate to think of the number of crackers we blew up, some in very inventive ways to enhance the boom. Every one of those explosions ended up with a pile of rubbish. Disorder came from every explosion, I witnessed nothing forming from all those explosions. Not even a pattern from shredded cracker paper casings. It was all chaos and destruction.

I've never heard of anything being created from huge bomb explosions. The atomic bombs dropped on Japan were among the biggest explosions witnessed by mankind. It obliterated and killed everything, contaminating the land so nothing would grow. No new life came from the explosion. It created no new cells or life forms. Order and new life don't come from disorder, that's a law of life.

Evolution scientists' theory is opposite to the law of life; however, they claim this cataclysmic event started it all and they refer to it as:

The Big Bang

They have also changed their story of how it came about many times over the years as they continue to conjure up new versions they think are more plausible. The latest one I'm aware of is that the Big Bang wasn't an explosion in space, it was actually the appearance of space. With a big bang the universe expanded from nothing into an extremely hot, very dense single point in space, then becoming everything we see today. It just mysteriously happened, something from nothing.

Theoretical Scientists, the guys who spend all day dreaming up this stuff, claim it was a monumental explosion or growth spurt that happened in something like a hundredth of a billionth of a trillionth of a trillionth of a second and this created a super high temperature that then cooled to just the right temperature. This in itself is an amazing miracle, as optimal temperatures are so vital for formation of new life. From a supercharged heat, it kept cooling until it reached the perfect point for life to form. Then, somehow the stop button was pushed, and it has remained in that state ever since.

Do you find this too extreme to believe?

Remember, this is just theory stuff, they have no actual proof to back it up.

Quarks and electrons formed, then protons and neutrons combined to form nuclei, thus making the first atoms; all the stuff required for life just miraculously formed. These things don't form easily, all the environmental conditions have to be perfect, and the odds of all this happening just by random chance from an original explosion is science fiction. There isn't a mathematical equation big enough to express the odds of it happening the way they claim.

The bottom line is, there is no proof of what happened and how it happened, no one witnessed the event, it's not recorded, nor has it been replicated to prove their theory.

If the foundation of evolution, the Big Bang, can't be proven then nothing else thereafter is credible, including evolution.

You can't construct a building on a faulty foundation. It will eventually fall over, and this is exactly what happens with evolution. The foundation of their entire theory can't be proven, which explains why the evolution scientists are constantly forced to invent new theories as they are desperate to prove it happened.

In the next chapter, I cover their Big Bang dating theories and how they come up with the 13.7 billion year number for the date of creation.

The theory of how the solar system formed really takes some incredible faith to believe. Supposedly gravity drew dust and gas together to create the solar system, with the sun forming first.

Waves of energy traveling through space compressed clouds of particles together, then gravity caused them to collapse in on themselves and start to spin.

This spinning caused the cloud to flatten into a disk like a pancake. In the center, the material clumped together to form the sun.

To summarize, they claim the sun was born from clouds which were compressed over a period of 50 million years, creating pressure which started the fusion of hydrogen that fuels the sun today. Apparently, the core of the sun can reach 27 million degrees Fahrenheit, this huge fireball was created just from compressed clouds. I'd like to see them prove that in a laboratory.

I'm not sure how gravity would pull this immense volume of dust and cloud into a single point because in space there is very little gravity. We see how astronauts float around when they are outside the space station. If their lifeline breaks, they float in space forever. Same with the junk floating in space, old discarded satellites float in space forever. They don't get compressed into a big ball of junk, there isn't enough gravity in space to do that.

With very little gravity in space, it's amazing there eventually became enough pressure to compress all the clouds. It would need an unfathomable volume of clouds and a huge amount of pressure to create this gigantic fireball 864,400 miles across, 109 times bigger than earth and weighing 333,000 times as much as earth. That's a lot of high force compression.

I would also love to know how they explain how all the planets were hung in space. Do they have proof of how they ended up in this perfect place and how the invisible path they follow was created? This is especially true for earth since if it were just a little bit closer to the sun the radiation and solar wind would "blow off" our atmosphere.

A little bit further away and we would be like Mars, too cold. Interestingly, in their search to find other earth-like planets they are now referring to where our planet is located as the "Goldilocks zone."

Also, consider the tilt of the earth which allows for four seasons. Then throw in the moon, which the earth needs to maintain this tilt. Otherwise we would wobble like a basketball with the north pole at times facing straight on to the sun, causing mankind to be constantly migrating as our polar ice caps would be all over the place as this wobble occurred. It is also interesting how the moon is the exact size to cover the sun for an eclipse and how it shields us from many hostile objects, as a view of the dark side of the moon reveals. The moon's gravitational pull pulls the seas towards it, causing ocean tides, which helps balance the global temperature.

Once again how the sun formed is just another theory. If the sun was born billions of years ago no human was around to witness it, and science hasn't recreated another sun to prove their theory is true. This stuff comes from astronomers studying the myriad stars in the Milky Way and making up models to arrive at a theory.

Evolution scientists make statements phrased in a way so as to mislead the public into thinking this stuff is true and proven scientific fact. It's not, it's nothing more than ideas and theories, and they know they will never have to prove them. It's pure science fiction. Even processes they claim date artefacts and rocks are highly flawed, as you will read in a subsequent chapter.

As scientists have no proof, they are desperate to find something of substance. The big goal is to prove the Big Bang, their holy grail. As it can't be proven, their entire evolution theory goes up in smoke.

Well, they are actually trying to do just that; spending billions on a machine called the 'Hadron Collider' designed to create a Big Bang model.

The problem with this massively expensive experiment is there is no atmospheric similarity between today and 13.7 billion years ago, as they don't know what it was. They have created a vastly different controlled artificial environment to suit their experiments. Therefore, any results they achieve will be flawed.

Scientists are now playing at being a creator as they are orchestrating the event. But this isn't what they explain evolution is about. They claim it happened naturally with no supreme being or intelligent force behind it, yet they are trying to create something that only happened once in 14 billion years. Or could it be they finally believe how impossible it was and are trying to show how a Creator created the universe.

Ponder this; they claim the universe was created from a Big Bang. Please explain how over the 13.7 billion years since the bang there hasn't been another Big Bang. Isn't it strange that this only happened once in 13.7 billion years?

If all the conditions were right for one explosion, why hasn't it occurred again, either before or after.

Hang on, it's actually longer, the 13.7 billion year number is since the bang. How many years since the beginning of 'whatever' originally existed before The Big Bang? They can't claim there was nothing as that would say before the bang nothing ever existed, and in that case, we can't be here today. You can't get something from nothing, that's a scientific fact and everyone knows that. There had to be something to produce the bang.

So, the Big Bang could have started it. Once again, their science fiction theory self-destructs.

To continue the evolution theory, it has to go back to a beginning. The Big Bang couldn't be the beginning, something had to exist prior, so the Bang had something to ignite. You can't have a bomb unless you have the necessary ingredients and environment for the bomb to explode. So, what was here before the Big Bang, where did the material come from to fuel the bang? It's sounding like it's pointing to a Creator.

Scientists are so obsessed with proving the Big Bang happened they have embarked on an attempt to recreate the beginning of life and are spending an obscene amount of money on their Big Bang theory proving toy, the 'Hadron Collider.'

Hadron collider

On the border of France and Switzerland, 300 odd feet underground, scientists have created a monumental and expensive toy to try and recreate the big Bang. They are wasting billions, yes, multiple billions on this machine, it eats big chunks of money every day. They built a 27-kilometre ring underground tunnel to house the 'Large Hadron Collider'.

The collider is operated by 'CERN', the European Organization for Nuclear Research, near Geneva, Switzerland and funded by;

Austria, Belgium, Bulgaria, Republic, Denmark, Finland, France, Germany, Greece, Hungary, Israel, Italy, Netherlands, Norway, Poland, Portugal, Romania, Serbia, Slovakia, Spain, Sweden, Switzerland and United Kingdom.

Cyprus and Slovenia are Associate Member States in the pre-stage to Membership.

Croatia, India, Lithuania, Pakistan, Turkey and Ukraine are Associate Member States.

The following countries have observer status; Japan, Russian Federation, United States of America, European Union, JINR and UNESCO.

If Americans think so what, it's not our tax money so let them have their fun, sorry, you're also getting in on the act. In a Science Jan 17, 2020 magazine, they report:

The United States has taken a key step toward building its first new particle collider in decades. Last week, the U.S. Department of Energy (DOE) announced that the Electron-Ion Collider will be built at Brookhaven National Laboratory in Upton, New York. The machine would enable nuclear physicists to probe the mysterious structure of the proton and how its mass and spin emerge from a teeming sea of even smaller subatomic particles inside it. The collider will cost between $1.6 billion and $2.6 billion and could begin to run by 2030.

(Science 17 Jan 2020: Vol. 367, Issue 6475, pp. 235-236 DOI: 0.1126/science.367.6475.235)

To justify spending billions of dollars of taxpayer money 'CERN' claims their work is to:

"uncover what the universe is made of and how it works. We do this by providing a unique range of particle accelerator facilities to researchers, to advance the boundaries of human knowledge."

Their particle accelerator is a very, very, very... expensive toy, the world's largest and most powerful particle accelerator with a 27-

kilometre ring of superconducting magnets and a number of accelerating structures to boost the energy of the particles along the way.

What is a **particle accelerator**? It's a big tube tunnel that fires two high-energy particle beams *(high-energy beam of atomic or subatomic particles)* traveling at close to the speed of light before colliding with each other. The purpose is to create an explosion so they can claim the 'Big Bang' actually happened? The speed of light is 186,282 miles per second (299,722 KM/sec). This is the stuff these guys waste billions on. Imagine the real good they could do if they took up medicine and spent those billions finding cures for diseases that kill millions of people each year, children in particular. If they did the world would be a better place.

This is one serious toy. The tubes are ultrahigh vacuum tubes and the beams are guided around the tube by a super strong magnetic force driven by superconducting electromagnets that need to be chilled to -271c, that's colder than outer space. To achieve this, they built the world largest 'Cryogenics' cooling system to produce the extremely low temperatures required. It's one of the coldest places on Earth. The cryogenic system requires 40,000 leak-tight pipe seals and 40 MW of electricity, which is 10 times more than is needed to power a locomotive, and 120 tonnes of helium to keep the magnets at -271c.

The project started in 2008 with an initial estimated cost of 10 billion, but to date they have blown over 14 billion on this monstrosity. Money generously provided by member country's politicians. That should make you feel good knowing your tax dollars are funding a gigantic scientific toy looking at trying to recreate the big bang.

While 14 Billion is the official figure, there are other reports claiming the true cost is over 50 billion when taking into account the cost of the thousands of staff paid by member countries. The numbers are obscene.

To make you feel better about your tax dollars being plundered, look on the bright side, you have invested in an impressive machine.

- It has 104 kilometres of piping under the vacuum, the world's largest.
- The insulating vacuum, equivalent to some 10^{-6} mbar, comprises 50 km of piping, with a combined volume of 15,000 cubic metres,
- Building the vacuum system required over 250,000 welded joints and 18,000 vacuum seals.
- The remaining 54 km of pipes under the vacuum are the beam pipes, through which the LHC's two beams travel. The pressure in these pipes is in the order of 10^{-10} to 10^{-11} mbar, a vacuum almost as rarefied as that found on the surface of the Moon.
- To direct the beams around the accelerator it uses 1232 dipole magnets 15 metres in length which bend the beams, 392 quadrupole magnets, each 5–7 metres long, which focus the beams.
- Just prior to collision, another type of magnet is used to "squeeze" the particles closer together to increase the chances of collisions. The particles are so tiny that the task of making them collide is akin to firing two needles 10 kilometres apart with such precision that they meet halfway.
- The Magnets produce a magnetic field of 8.33 tesla to keep the particle beams on course. A current of 11,850 amps in the magnet coils is needed to reach magnetic fields of this amplitude.

- It achieved the hottest manmade temperatures ever, 9.9 trillion degrees Fahrenheit, by colliding lead ions as they try to replicate the Big Bang.

I trust knowing all this made you feel much better.

It was first turned on in August 2008, then stopped for repairs from September until November 2009. Its moment of glory was their claim they **THINK** they found Higgs boson, that's what you get for your 14 - 50 billion. Just a **'THINK WE FOUND'**. That what you get for 50 billion.

What is a Higgs boson?

The Higgs boson (or Higgs particle) is a particle in the Standard Model of physics, named after Peter Higgs, the first person to suggest it might exist in the 1960s.

They claim a Higgs particle is thought to be the particle responsible for all physical forces. A boson is a subatomic particle, such as a photon which has zero integral spin and follows the statistical description given by S. N. Bose and Einstein.

Higgs boson is the quantum particle associated with the Higgs Field, which gives mass only to elementary particles such as electrons and quarks. Quarks form the protons and neutrons in atomic nuclei.

The Higgs field, they claim, is of utmost importance to particle physics theory, so this gross extravagance and waste of money is just to prove a theory. It won't help saving lives or cure people from diseases or anything worthwhile, but it will make scientists feel good.

Scientists claim it's a big deal as it is very difficult to detect the Higgs boson because it doesn't stay around for long. Basically it decays instantly, and normally there aren't any around because of the amount of energy required to make one.

In simple English, we have thousands of scientists who have spent 50 billion making a big toy to find a particle that's nearly impossible to see, and even if by some chance it happens to be created, it decays instantly so you can't see it. You have nothing left to show anyone, and it won't improve the life of anyone.

The Large Hadron Collider at CERN was built mainly for this reason. They claim there is a one-in-10 billion chance of one of these particles appearing and being detected by the bunch of supercomputers sifting through huge amounts of data. It can't be seen, and it can't be bottled to observe.

The insanity gets better, here's the good part.

In 2011 they announced results which **COULD** suggest the Higgs boson existed, but they didn't know if it was true.

In 2012 they announced they had discovered a particle they **THINK** is a Higgs boson. This came from the combined effort of thousands of physicists, engineers and technicians, an enormous payroll, and all we got was a **THINK!** No doubt they had to say something to justify the colossal waste of money and use to entice gullible irresponsible governments into throwing more money at them.

In 2013 they did more testing and announced that:

"New results **INDICATE** *that particle discovered at CERN is a Higgs boson' and results further 'elucidate' (their word in 14 March press release), the particle discovered last year"*.

Elucidate means; *make something clear or explain.*

After another year's work you get an 'elucidate', they think they found it. Sorry folks, that's all you get for your 50 billion.

No doubt for scientists this is fun stuff, and they would be aghast at my cynicism so they would go on to invent a very long waffling scientific paper justifying their obscene waste of money.

Who are these guys accountable to, who keeps paying their bills when there is so much need in the world for real scientific discoveries to improve people's way of life, and cure illnesses?

If you're wondering why I explain all this, here's the reason.

Higgs boson has been labelled "The god Particle' by Peter Higgs, the first person to suggest Higgs boson might exit.

In 2012 when CERN announced they THINK they found it; the media went crazy, referring to it as the *'the god particle'* along with claiming it's what started the Big Bang.

Truth is, this has nothing to do with it and the media totally misled the public. The name 'god particle' apparently came from the story of Nobel Prize-winning physicist Leon Lederman who referred to the Higgs as the 'Goddamn Particle'. The nickname was meant to poke fun at how difficult it is to detect and identify, as previously mentioned.

'The Goddamn Particle' was supposed to be the title of Lederman's book published in the 1990s about the history of particle physics. His publishers didn't like it, so they changed the title to *The God Particle*, hoping to sell more books.

They still haven't found the magical particle and there is still NO EVIDENCE the Big Bang happened. They should rename it the *50 billion particle.*

Experiments surrounding Higgs boson apparently show the particle would have a mass of 125 billion electron-volts, or more than 130 times the mass of the proton. This then led to another problem, at that mass Higgs boson should have destroyed the universe just after the Big Bang as Higgs particles theoretically attract each other at high energies. This they can never prove as the energies must be extraordinarily high, *'at least a million times higher than the Large Hadron Collider is capable of attaining'*, according to Arttu Rajantie, a theoretical physicist at Imperial College London.

They claim just after the Big Bang there was enough energy to make Higgs bosons attract each other. This could have led the early universe to contract instead of expanding, thus annihilating it just after it happened. If that was the case, the result of the Big Bang, if it ever happened, may have left nothing. Add that one to all the other impossible factors that had to come together by random chance.

As you can see, there are many unknown factors and conflicting theories. Bottom line is there is no scientific evidence or facts to support the Big Bang theory, it's just conjecture. The same as the theory of the green Martians coming out of another dimension in a black hole and creating the earth.

Even if they discovered a Higgs boson, it wouldn't prove the Big Bang happened or how the universe was created or when, as the experiment is being done in a totally different controlled environment.

Theoretical physicists

CERN employs many Theoretical Physicists. These are people who sit around most of the day theorizing nature, dreaming up ideas that may or may not be true and provable. If enough of their own kind agree it's a good idea it goes on to become a theory, and they go as far as assuming that since they all agree this makes it true. This is where these crazy theories come from. Lesson learnt, don't believe a theory. It's not science or fact, it's just an unproven idea.

The fictional character Sheldon Cooper of the comedy series *The Big Bang Theory* played the role of a Theoretical Physicist. Although a comedy and very entertaining, I was amused at the total lack of common sense he acted out, obviously there was a basis to it in order for it to be funny.

Below is CERN's own explanation of Theoretical Physicists from their website (2020).

Theoretical physicists are rather typical scientists. If you imagine them as absent-minded, egg-headed, bizarre characters scratching their chins while deeply engaged in thought... Well, most of the time you'd be right.

What these people do is to try to figure out how Nature works. That is, why the stars shine, why water is fluid and the sky is blue, what you are made of and why does "it" weigh that much, why the universe expands, or what energy and matter actually are...

The main specialty of theoretical physicists at CERN is trying to understand "elementary particles", which are the fundamental constituents of the Universe and the agents of the basic forces of Nature, like gravity. As it turns out, our ever-advancing knowledge of these "elementary" little things is also the basis of our understanding of the Universe as a whole!

If you made a short visit to CERN's Theory Division, you might think that you are in a zoo. But that is not *entirely* right. True enough, you will find women and men of dozens of nationalities, cultures, languages and what not... some of them may even look like ET. But what these people are doing is what defines our species in its ensemble: asking, and sometimes answering, some of the deepest questions. Thus, the zoo is more like a circus of magicians, in which the performers - uncharacteristically - would insist in showing you their cards... and the entrance is free

By their own admission, they admit they are a bunch of clowns. Yet, these are the people irresponsible politician's give billions to so they can play with the Hadron Collider and sit around all day concocting crazy ideas. And to think evolutionists claim humans are becoming more intelligent and the world is becoming a better place.

Scientists working to prove evolution are so desperate they are even trying to change the definition of science to suit their unproven ideas. This is necessary if they hope to con the public by claiming a theory is proven science.

Atheist Professor Richard Dawkins made this very disturbing statement:

"Science proceeds on intuitive leaps of the imagination building an idea of what might be true".

WOW, this is what evolution scientists have degraded to. According to Dawkins, science now includes the imagination of the researching scientist. If scientists can imagine something happening and think it might be true, according to them then it's defined as a scientific discovery and it is true.

This is one reason I wrote this book; theoretical and evolutionary scientists have deviated so far from scientific truth they have lost all credibility.

Naturally I'm not referring to productive scientists who work hard to come up with proven verifiable discoveries like those in the area of medicine. Those guys are great people who help humanity while doing really worthwhile positive work. They should get the $50 billion.

Yes, you read it correctly, they can't prove evolution, so they claim 'imagination' is the basis for science. That's not science, it just someone's idea.

They are making assumptions, inventing concepts of objects that can't be seen or proven, and then claim this is science. *It's certainly not science.*

I was taught never to assume anything; it means you're too dumb or lazy to take the time to find out if it's true. Something is either true or it's not.

We can no longer take for granted what these scientists claim is true, we need to question any scientific finding to clarify how much is "imagination" and how much is proven quantifiable scientific fact.

Perhaps the new norm is a group of like-minded individuals with a common interest throwing around a bunch of ideas, and if they agree it sounds plausible, it goes on to become a scientific discovery. That's what Professor Dawkins is saying science is.

Here's an idea. If enough of us agree on the below theory and think it could be plausible, we can claim its science, then we can announce we have discovered how the earth was made.

How does this theory sound?

Martians from planet Zorb entered this universe through a black hole (Science doesn't know what's in a black hole, there could be other universes for all we know). From the cargo area of their spaceships the Martians released all the living cells with the designed DNA code necessary to form the earth. They did this to replicate their planet in the black hole.

After finishing their initial work, they left behind workers to finish the design of earth and animals. We can't see them as they are invisible, but they are the intelligence behind evolution. Well, it's actually not evolution as that's impossible. It's creation, but they have structured it so silly earth scientists think its evolution.

The spaceship commander didn't want to hang around for billions of years, well actually in their realm there is no time as they are from a black hole where time doesn't exist, so they don't know when to come back again, which is why we haven't seen them. One day they will appear from one of those black holes and take over the earth. In the meantime, unobserved highly intelligent worker Martians have directed the creation of earth and everything in it, and here we are today. Thank you Zorb Martians.

This concept is such a good idea we have named it "The theory of Martianlution". If I had time to waste, I could write an entire book on our new theory, filling it with lots of ideas from my imagination that would make a plausible argument for it. Then, if you all liked the sound of it, we would now have a new scientific discovery. Stupid yes, but a fitting analogy for what evolution scientists claim is their version of truth.

Despite scientists wasting billions of dollars trying to prove the Big Bang in artificial conditions, any claim they make will be irrelevant for it is built on flawed premises. They cannot replicate the actual environment and conditions present at the time of their alleged Big Bang. They can't replicate creation of the earth and universe, and everything in it. Whatever they produce still won't prove the Big Bang happened for it is impossible to recreate the original conditions on which their theory is based.

It won't pass the accepted scientific process for proving a theory is true and it never can. It hasn't been replicated, repeated, or witnessed. It's only a theory.

There are peer-reviewed papers from the scientific community questioning the calculations behind the Big Bang theory. Sectors of the scientific community also agree the science isn't there to prove it.

In spite of its mathematical simplicity and observational triumphs, the Big Bang model of the Universe remains an unfinished work of art. ***Many of its late-time successes can be traced to the initial conditions postulated for its early stages, and these are put in by hand, without justification, other than to retrofit the data.*** *The main culprit for this shortcoming is the so-called horizon problem: the cosmological structures we observe today span scales that lay outside the ever shrinking "horizons" of physical contact that plagued the early universe. This precludes a causal explanation for their initial conditions.* (Niayesh Afshordi and João Magueijo, Theoretical Physicists paper 'The critical geometry of a thermal Big Bang')

Note they use **'*postulated*'**, meaning; assumed, suggested, and *'put in by hand without justification other than to make data fit'*.

Confirmation scientists use information they reverse engineer to arrive at the answer they want. This is called lying, deception, and manipulating evidence. People go to prison for that. I don't know what *'observational triumphs'* are as no one has observed the big bang.

Lesson learnt, never believe anything you hear, or claims made by evolution scientists because they rig the information.

Conclusion - Science doesn't have any real proof of the Big Bang, it's all theory and they haven't and cannot prove evolution through the application of scientific principles. Therefore, they cannot claim evolution is an absolute truth.

Evolution is not science. It's an idea, a theory in the minds of individuals wanting to convince the public evolution is true for it fits their worldview of rejecting a Creator and promoting their religion of atheism.

Many scientists who believe in evolution are atheists, such as Richard Dawkins. Their opinions and thoughts are based on their worldview, which is atheism.

Atheism is a religion, and according to the dictionary definition of 'religion' so is evolution.

Religion

faith, belief, teaching, doctrine, theology; sect, cult, religious group, faith

community, body, following, persuasion, affiliation.

Evolutionists don't follow science; they follow their atheist religion of evolution which is designed to deny a Creator and further their religion.

How reliable are dating methods for the Big Bang and evolution?

If evolutionists claim the universe is 13.7 billion years old and earth's physical geography is millions of years old as they claim on all the nature programs, are these dates actually based on proven science?

Do we naively accept the claims of evolution scientists as truth when there are so many examples of them quoting theories as proven fact when they are not?

What processes do they use to arrive at these dates? Are they reliable, and what is the science and methodology behind dating? Is it fact or based more on their assumptions?

I started this chapter intending to include an overview of the cosmic calendar which breaks down the history of the universe based on their dating techniques and compressing it into a single year.

They then list key events at various places on the calendar with man only appearing in the last few seconds of the last day.

However, I faced a major problem. There are endless opinions and theories about events, and it would add an extra 100 pages to the book, turning it into another boring volume consisting of hypothetical theoretical ramblings. At the end of this new revised book you wouldn't be any wiser as they are all different theories and viewpoints with one thing in common, they have no credibility.

If you're interested in reading all the endless theories, you can Google them. Just remember, despite being written as if true and a scientific fact, don't be deceived by their rhetoric. Watch for the key phrases, "we think, it appears, we have reason to believe, etc." for these are the buzzwords used to convince you they are the experts in this area.

There are some crazy theories floating around in evolutionists' Fantasyland. Fact is there is absolutely no verifiable evidence on dating, it's all theory and based on flawed assumptions, which you are about to discover.

My solution for tackling this issue of dating is to summarize the main claims. I realize and fully expect it to be challenged as not being accurate, and I'm not claiming it is. That would be impossible as there are so many different versions and theories, and the list keeps getting bigger. If I do somehow get it wrong, the evolutionists have no basis to complain based on their own rules as it is just another theory, and none of them can claim which one is actually the correct one.

First, some history and evidence illustrating why the dating techniques for evolution is flawed. Despite the vast periods of time they place on events and items, they don't actually know when something occurred, it's all a big guess based on assumptions.

When evolutionists are challenged on the reliability of their dating processes a common answer is, *'we are now better at working it out than we used to be'*. That gives a lot of assurance on the reliability of their previous methods. It's also saying in 50 years they will have different dates, claiming they have gotten even better at working it out. This in itself is an admission the dates they throw around are just theory. They don't really know but they are still trying to present their dating results as fully verifiable.

History proves my statement. In Darwin's time scientists said the earth was 100 million years old. Then, forty-odd years later they told us it increased to 1.6 billion years old. A few years later they again revised their numbers, this time doubling it, then about 40 years ago we were told they finally settled on a number; it's really 4.6 billion years old.

Should we believe them this time since based on their history, they will eventually revise it again, claiming they have developed even better ways of calculating the age of our planet. Who knows, in another 40 years they could announce they have discovered the earth is 9 billion years old.

If they claim the universe is 13.7 billion years old, where does this come from, how do they arrive at that date? Short answer is, it's all a theory, there is absolutely no irrefutable evidence the universe is 13.7 billion years old. The end.

They theorize it all started at a single point when the universe went through a super-fast inflation from a tiny atom expanding to the size of a grapefruit in a fraction of a second. No one can explain where the atom came from if nothing existed before the bang.

Then it went into a 'Post-inflation' phase as it's now just a seething hot soup of electrons and other particles. No one knows or can prove how they formed. Remember, all this happened in just a fraction of a second. Whilst the soup is still too hot to form atoms, charged electrons and protons prevents light from shining, making the universe a super-hot fog, much like our brains trying to understand all this. I don't know what light source the fog supposedly blocks as at this stage the sun doesn't exist. They claim all this happens in just 3 minutes. How would they know that other than by making another assumption!

This muck quickly cooled to just the right temperature so the building blocks of matter could magically form, including the quarks and electrons of which we are all made. Another miracle then occurred. A few millionths of a second later, without any control, designer or planner, this cosmic goo magically cooled to exactly the right temperature necessary for life to begin. That's quick, quarks aggregated to produce protons and neutrons. Within minutes, these protons and neutrons combined into nuclei. As the universe continued to expand and cool, which is only speculation, things began happening more slowly. After all that happened in these first few minutes, it took 380,000 years for electrons to begin orbiting around nuclei, forming the first atoms. And yes, they claim this just all miraculously happened by itself.

There is absolutely no proof of all this, it is nothing more than a theory invented by scientists who talk as if they proved these events actually happened as they claim. It is here where evolution scientists deceive people.

Then over about another 300,000 years, I don't know what years they are referring to as our calendar wasn't formed yet. We base an earth year on the period it takes for the earth to complete a single orbit around the sun. Do you see a problem here? These initial dates occurred during a period when there was no sun or planets, including the earth. They hadn't evolved yet so there was no basis for recording time. Time didn't exist, no sun, no time. So how can they measure years before the creation of time. This alone throws out any of the years they quote. It may have all happened much faster than their theories suggest.

Perhaps they are using Martian years!

So, at the end of the 300,000 years electrons combined with protons and neutrons to form atoms, mostly hydrogen and helium, allowing light to finally shine. Must have been Martian light from their spaceships as no sun, moon, stars or planets existed.

Basically, it all started with a single atom which nuked itself, then 300,000 years later we have another atom!

Don't worry if you're confused, it's the normal state of mind for those trying to understand evolution so you're not alone. Perhaps the problem is we haven't yet evolved to the same state as the evolutionists, who claim they understand and believe this stuff even if it's more science fiction than science fact.

After a billion years, gravity causes all this hydrogen and helium gas to form giant clouds that go on to form galaxies while smaller clumps of gas collapse to create the first stars. They claim all these gases get together to form solid structures, the stars, just by random chance. Yes, the first stars formed by collapsing gases, with the force of gravity compressing clouds to form all the stars and galaxies we now see. I'm not sure how that could happen because, as you read in the last chapter there is very little gravity in space so how could there be enough gravitational force to achieve this monumental feat.

I would like to see a mathematical equation for the odds of that happening. For just one star forming it would be an ENORMOUS number. It gets even more mind-boggling as they claim it was repeated for the estimated 100 billion stars per galaxy. This means had to happen 1,000,000,000,000,000,000,000 (that's 1 billion trillion) times, as that's the estimated number of stars just in the observable universe, what are the odds of that happening, ZERO.

It gets even more mind-boggling. Over the course of these 15 billion years the galaxies began clustering together under gravity. Remembering space has little gravity, the first stars die, spewing heavy elements into space. Yes, despite claiming the universe is 13.7 billion years old, somehow, the age now went up to 15 billion.

Those old stars that died eventually turn into new stars and planets, bringing us to the present generation. All this has been done miraculously with no intelligent force, no planning or design, just by the invisible magical force of evolution. May the force be with us.

What are the odds of the sun and moon, which the earth needs to exist, forming from nothing? The sun is a ball of nuclear fusion capable of reaching a temperature of 27 million degrees Fahrenheit. It's hung in space at just the right location about 93 million miles from earth. If much closer earth would fry or further away earth would be a big ice block with no life. It had to be exactly where it is in what evolutionists are now calling the "Goldilocks zone" to create and sustain life.

And they tell us all this just happened to occur as the result of a freak accident, that somehow the sun hangs in space at just the right distance from earth.

Here's another problem for evolutionists. They claim formation of the stars started around 1.6 million years after the big bang event when the environment was in just the perfect state for this to happen. They claim the universe was in a continual state of change before arriving at that state, then miraculously, at just the right stage, all these bizarre changes stopped as if someone turned off the switch, and things have remained the same ever since. Of course, there is zero evidence of how that happened.

If, as they claim, these elements continued to change, the environment would have continued changing and the balance of gases and clouds would also have changed, thereby making it impossible for stars to keep forming.

They constantly contradict themselves. At a certain point gas in the universe goes through a prolonged process of change then somehow the brakes go on at just a certain point in the overall 14-billion-year age of the universe. Then they claim the mixture of gases has stayed the same ever since. Why? If these changes occurred for billions of years, what caused them to suddenly stop at just the perfect time for stuff to form. Must have simply been a miraculous coincidence; either that or the Martians arrived with fire extinguishers to cool things down. Anything is possible as it's just science fiction.

Cosmologists and evolutionists still claim the universe is continuing to expand, with new stars and galaxies forming so the balance of gases continues to remain perfectly balanced. After billions of years of change, at a certain point when the brakes went on, it all stopped at that point and nothing has changed since. That's their theory.

There is still another problem. After their special 1.6 million point in time, they also claim the atmosphere kept changing so the first organisms were able to form. Isn't that nice of them? This was the start of life and all that allegedly happened around 3.8 billion years ago. This means the composition of gases was still changing. If at this poi nt you're really perplexed you're not alone, so am I. But don't worry, no matter how hard you try you'll never be able to make sense out of all this as none of it is truth, it's just science fiction theory.

If an evolutionist is reading this and willing to be honest, they can't disagree with what I have written, as in the domain of evolution there is no truth. Their entire Big Bang and universe evolution fairy tale for grown-ups is based on assumptions and theories, so truth is irrelevant. Besides, there are so many theories, even they probably don't know which version I'm referring to.

How did they invent the 13.7-billion-year date for the Big Bang?

In 2013 they measured the distances and radial velocities of other galaxies, most of which are flying away from our galaxy at speeds proportional to their distances, called the expansion rate. Using this estimated current expansion rate of the universe, they then roll things back to the point where everything was supposedly contained in a singularity. I.e. a one-dimensional point containing a huge mass, a point of infinite density and gravity. Before this event, some claim space and time didn't exist. They then calculate how much time must have passed between that moment (the Big Bang) and the present.

At face value this seems fairly logical for it is based on a mathematical calculation, but the equation only works based on the assumption their expansion rate calculations have been consistent and unchanging since time began. There is no way to know for sure if this assumption is correct or not. Therefore, the date they arrived at is based not on fact but assumptions.

The unknown factor is the history of the expansion rate. Has it been constant, meaning things have been expanding at the exact same rate for the billions of years they arrive at? They estimate this date by examining the current density and composition of the universe using observations of the cosmic microwave background. This is the electromagnetic radiation left over from an early age of the universe known as 'relic radiation', which is allegedly left over from the Big Bang, which they can't prove actually happened. Their expansion rate isn't based on evidence, it's another assumption more or less based on circular reasoning.

They performed their calculations using the Wilkinson Microwave Anisotropy Probe (WMAP) and Planck. In physics, the Planck length is a unit of measurement consisting of the distance light travels in one unit of Planck time. Using this standard or measurement they came up with 13.77 billion years, give or take 59 million years, but what's a few million years among evolutionists.

Scientists claim they can see objects 13 billion light years away from earth using the Hubble space telescope. We don't know if that's true or just some fancy calculation these guys frequently invent to suit their narrative. To give a perspective on how far that is, if you could travel at the speed of light you would be able to go around the earth 7.5 times in just one second.

Hence my skepticism if they can actually see something 13 billion light years away.

Once again, the 13.77 billion calculation is full of assumptions. At one point they thought all the galaxies must have been closer together and packed together in a tiny point and the expansion rate was constant, now they think it's speeding up. I don't know how this could work because according to what we read, they claim the Big Bang was an immediate super-fast monumental explosion or growth spurt that happened in something like a hundredth of a billionth of a trillionth of a trillionth of a second. If that was true, how can they possibly estimate the expansion rate. If you follow their line of thought, at that force and velocity the universe could have rolled out to its present size in a very, very, very, very short space of time. What a coincidence that this is what Creationists claim.

Once again, there are so many unknowns, so many variables they don't know about, yet despite this lack of information they claim they have proven the universe is 13.77 billion years old. As we know in the evolution scientific community, the word 'proven' actually means we have a theory we think is true.

All this came from a timeline chart prepared by one of those theoretical scientists, the ones CERN call their circus of performers.

It's just one giant science fiction theory, things blowing up in space and randomly forming all our planets. The sun, moon and earth fixed in just the right place in space so earth can exist in a pristine state for life, and all these planets formed from a bunch of gases compressing in a low gravity environment, and all just by random chance.

So, where did the Big Bang stuff originally come from; it goes something like this:

It all started because of deep-space observations conducted in 1912 when American astronomer Vesto Slipher measured the first 'Doppler Redshift' of a 'spiral nebula' which is the obsolete term for spiral galaxies. In physics, redshift is a phenomenon where electromagnetic radiation (such as light) from an object undergoes an increase in wavelength. Examples of redshifting are a gamma ray perceived as an X-ray, or initially visible light perceived as radio waves.

If a source of light is moving away from the observer, redshift appears. Objects move apart (or closer together) in space. This is an example of the Doppler effect. In almost all cases, spiral galaxies, the stars you observe with a spiral of gases surrounding them, were all observed moving away from our own.

In 1922, Russian cosmologist Alexander Friedmann developed what's known as the Friedmann equation, claiming the universe was likely in a state of expansion similar to a balloon being filled with air.

In 1924, Edwin Hubble's measurement of the distance to the nearest spiral nebula showed these systems were not individual stars but other galaxies. Hubble began developing a series of distance indicators using the 100-inch (2.5 m) Hooker telescope.

In 1929, Hubble discovered a correlation between distance and recession velocity–which became known as Hubble's law.

In 1927, Georges Lemaitre, a Belgian physicist and Roman Catholic priest, independently derived the same results as Friedmann's equations.

He then proposed the inferred recession of the galaxies was occurring because of the expansion of the universe.

In 1931, Lemaitre took this further, suggesting the current expansion of the universe meant the further back in time one went, the smaller the universe would be. At some point in the past, he argued, the entire mass of the universe would have been concentrated into a single point from which the very fabric of space and time originated. This single point was where the Big Bang started. This was the start of the Big Bang theory.

The Big Bang was born.

Physicists then did what they do best, theorize. The majority supported a theory the universe was in a steady state and new matter was being continuously created as the universe expands, thus preserving the uniformity and density of matter over time. This theory was much like our 'Theory of Martianlution'.

Other theories also surfaced, and the debate raged on between the physicists trying to decide on the Steady-State Model or the Big Bang Theory.

Astronomer Fred Hoyle was a Big Bang theory supporter and the person who came up with the phrase "Big Bang" during a BBC Radio broadcast in March 1949.

Eventually, the Big Bang had more supporters, thus winning the debate instead of the Steady State theory.

Physicists kept working on their beloved theory, including Stephen Hawkins. In 1981, physicist Alan Guth theorized the period during the rapid inflation phase mentioned above explained away one of their theoretical problems. In other words, one physicist found a hole in the theory, so another one solved the problem by inventing a different theory to plug the hole.

I trust you are seeing the pattern here; a bunch of physicist's theories invented the Big Bang.

There was still however a big black hole in the theory, Dark Matter. This is stuff that can't be seen or detected by instruments. I knew you are probably asking, if it can't be seen or measured how do they even know it exists? Good question.

They claim there's a mysterious unseen force that is causing the universe to continue to expand.

It's now sounding like Star Wars, 'The Dark Force', may the Dark Force be with us.

We might be right on with our theory of 'Martianlution' or perhaps there could be a Creator still at work, which evolutionists can't disprove.

Unlike normal matter, dark matter does not interact with the electromagnetic force. This means it does not absorb, reflect or emit light, making it extremely hard to find, hence the name 'Dark Matter'.

Researchers have only been able to infer the existence of dark matter based on the gravitational effects it appears to have on visible matter. It can't be seen, so infer is another word for theory in their language.

They claim the amount of dark matter seems to outweigh visible matter by a factor of roughly six to one, making up about 27% of the universe. So, 27% of the universe is there but they can't see it, that's interesting. Could this be where our green Martians live?

See, I told you we could invent a new theory, and science has now confirmed where they live. Our theory is gaining strength as we are building on it with ever more theories, exactly as the evolutionists do. Soon we will be able to claim we have proven it, as science has proven where they live. This is the exact same strategy evolutionists use.

Another thought that would really upset evolutionists is dark matter might even be the realm where spiritual beings live. You know, the ones evolutionists don't believe in.

What is dark matter? One of their ideas is that it could contain 'hypothesized particles', their terminology, not mine. Hypothesized means, 'a proposed explanation', or in simple English another idea or theory. Isn't it fun learning all the different words they use for theory.

Experiments are currently ongoing at the very expensive money burning Hadron Collider you read about in the previous chapter. Now this one will really make you laugh. CERN admits that despite spending all this money, they think that even if dark matter particles were light enough to be produced at the Hadron Collider, they would escape through the detectors unnoticed.

However, they would carry away energy and momentum so physicists could infer their existence from the amount of energy and momentum "missing" after a collision. So, if they somehow manage to find dark matter, it can't be seen, and it will escape as soon as it's formed. But don't worry, despite all this they will 'infer' its existence because it's not there!

Now here is the big punchline, the argument cosmologists, physicists, and all the other 'ists' who throw their ideas into the mixing pot make for how they arrived at the age of the universe is based on the observation that massive stellar objects, many light years distant, are slowly moving away from us.

Meaning the universe is still expanding and the calculation they use is taking this rate of expansion and working it backwards to arrive at 13.7 billion years ago as their basis for the Big Bang dating.

In simpler terms, it's saying they measure the distance between two cosmologically distant points, wait a certain period of time and then measure the distance again, thus obtaining the rate of expansion. It's a lot more complicated than this as they can't use a ruler since there isn't one long enough. Even if they could, it would take many generations of humans to observe it, so they use very complicated equations and formulas such as; Comoving distance, Transverse comoving distance, Angular diameter distance, Luminosity distance and Light-travel distance.

This means they are basing their entire Big Bang 13.7 billion dating on the accuracy of these equations and formulas that could be highly flawed since we are talking light year distances on a magnitude that no human has travelled. If there were to be even the slightest error it could throw the calculations off by billions of years.

As we know from chapter two, the extinction of species is in rapid decline. The earth isn't improving as evolution theory claims. Instead, the planet is degenerating, or, in simpler terms, wearing out, so it's possible that if the universe is still expanding the rate could be slowing due to degeneration. In other words, the expansion rate may have progressively slowed since the Bang and was never constant.

This is a possibility they can't disprove; therefore, their assumption is stuffed. In fact, the universe might be much younger if these expansion rate numbers are variable.

Once again, although evolution scientists convincingly spin the yarn it all started with a big bang 13.7 billion years ago, there is absolutely no quantifiable evidence supporting this. It's all based on theories, assumptions, equations and formulas contrived from theoretical Physicists from the zoo. Highly suspect, to say the least.

Evolutionists scoff at Creationists and the faith they require believing in intelligent design and a Creator. I submit the opposite is true. Being an evolutionist actually require more faith as they are putting their trust in the odds of random chance, continually changing theories, assumptions, inferred chemical reactions, assumed formulas, and all the other unproven things evolutionists invent. It hasn't been scientifically proven and never will be. It's impossible to prove according to the long-standing principles required for a scientific discovery. By contrast, Creation has stood the test of time for thousands of years. Big Bang and evolution haven't, its only appeared in roughly the last 200 years.

Creationists have proof of a Creator, it's scientific, it's proven, and it doesn't require any faith, they see the results of it every day.

Rock and Carbon dating.

Another evolutionist's tool used to convince you into believing evolution is a fact is their dating of rocks, mountains, bones and fossils.

They claim their dating methods prove evolution and the earth's age. Once again, they claim its science and it's true. Sadly for them, once again it's highly subjective and based on more assumptions.

We frequently hear the dates of rocks, geographical areas, animal or human remains. Watch any nature program and how they haphazardly throw around various millions of years with no reference or explanation on how they came up with the date. Evolutionists work on the basis if something is said often enough, eventually people will believe it's true.

So how do they date a hunk of rock, is it accurate and reliable?

The methods used to date rocks, fossils, bones and artefacts are not simple. They have many unknown factors and flawed assumptions, making them unverifiable and therefore speculative. It's a VERY complex issue and fundamental to the evolutionist's argument, therefore one would think when they make these claims they would ensure they are based on verifiable scientific processes.

You have probably heard the phrase 'Carbon Dating', many think this is the only method used for dating and the benchmark. If only it were that simple. There are various methods of dating and carbon dating is just one of those methods and it only works on living fossils, bones and wood.

Below is a condensed summary of the various methods, along with the reasons used and associated problems and flaws.

Radiometric dating, Radioactive dating, Radioisotope dating.

Used to date rocks and other objects. This method compares the amount of naturally occurring radioactive **isotopes** in the material to the amount of its decay products. The rate of decay is called 'Radioactive Decay'.

The big flaw in this process is they estimate the decay factor based on laboratory tests over the past roughly 100 years, assuming the decay rate has been consistent. There is a big difference between observing something for 100 years compared to 2 billion years.

Ponder this. The dating is based on the claim everything in the universe has remained constant for billions of years, something we know is rubbish. They base their entire argument on the reliability of dating on the assumption 100 years of decay is the same as it may have been 1 billion years ago when they have no idea of the state of earth back then. The atmosphere may have been entirely different, the sun hotter or colder, or a host of other factors might exist to change the rate of decay.

They also don't know what other contaminants were present over these millions of years which could also affect the results. One must admit there is a VERY BIG possibility their assumed rate of decay is wrong, thus putting in doubt the entire basis of their dating methods. Especially when we know the earth is degenerating. This is the HUGE problem with evolution and earth age claims, so much of it is based not on science but on assumptions. They make rash comments claiming it is true and science has proven it, however when you look deeper into the processes used one discovers it's all built on assumptions. It doesn't stack up.

Scientists know this, and if you search the internet there are many articles relating to reliability of decay rates with experiments still being undertaken. If they are sure they are correct, why would they still be conducting experiments to confirm the validity of the method used.

Geologists have taken newly formed rocks with a known age and found that Radiometric dating highly inflates the age, thus proving the method is unreliable. Geologists can't prove or verify the rate of decay they use as there are so many variables.

If the rate of decay is an assumption, there is no way of determining if, in fact, their claim a rock is 200 million years old is true because there hasn't been an observance process over the 200 million year history verifying the rate of decay. It is an assumption and they could have it very wrong, throwing their 200-million-year theory down the drain.

What are the facts proving its unreliability?

Decay rate is a big issue as some elements decay much quicker than others. Uranium isotopes have been found to be unstable and decay, which can affect other isotopes they come in contact with such as lead.

Another problem is Radioactive elements are unstable and are constantly trying to move to a stabler state by giving off radiation.

The universe is full of naturally occurring radioactive elements. These radioactive atoms are inherently unstable; over time, radioactive 'parent atoms' decay into stable 'daughter atoms' (e.g. Uranium converts to Lead) which is a different chemical element. These changes have resulted in the atoms now having a different number of protons and electrons, they call this 'Radioactive Decay'.

Meaning... there is no way of knowing what the original state of any piece of rock was, as the daughter atoms now in the rock were not present back when the rock was created. In other words, the composition of the rock has changed and there is no way of knowing what it changed from.

If they claim a rock is 500 million years old, how could they know what the environment was when that rock was initially formed. Or what it was subjected to during that huge time period, which means it's unknown what has changed in the atoms.

Every situation would be different; therefore, they would have no accurate way of knowing the amount of parent isotopes or the amount of daughter isotopes present at day one. They cannot answer that exact question, and without an answer there is no reason to prove their dating methods are correct. Once again, they are assuming. Of course, they will give a long explanation of how they have done the calculation, but irrespective of what they claim, they are guessing as to the original composition of any particular rock

An 'isotope' is a form of chemical element whose atomic nucleus contains a specific number of neutrons in addition to the number of protons that uniquely defines the element. The nuclei of most atoms contain neutrons as well as protons. Every chemical element has more than one isotope.

Isotopes all have the same chemical behavior, but unstable isotopes undergo spontaneous decay during which they emit radiation and achieve a stable state. This property of radioisotopes is used in the archaeological dating of artefacts. Once again, the rate of decay is estimated; there is no true historical data.

Examples of isotopes are; carbon-12 (the most common and accounts for almost 99% of naturally occurring carbon). Carbon-13, and carbon-14 are three isotopes of the element carbon with mass numbers 12, 13, and 14, respectively. The atomic number of carbon is 6, which means that every carbon atom has 6 protons, so the neutron numbers of these isotopes are 6, 7, and 8 respectively.

Geologists can't take just any hunk of rock, it must contain the parent radioisotopes which are rocks that came from molten lava. These are Uranium-238, Uranium-235, Thorium-232, Rubidium-40, Potassium-40 and Samarium-147. Remember this, as it makes dating claims by property owners and nature programs highly suspect and they were probably never actually dated.

To give an example of the 'parent-daughter' transformation, the parent Uranium-238 converts to Lead-206. This may sound simple in theory when measuring this stuff, however it's not. The difficulty lies in precisely measuring very small amounts of isotopes. No doubt you will see the impossibility of working out the amount of Uranium-238 that was originally in a sample by the amount of Lead-206 present. Remember, they use tiny amounts of isotopes so who knows how fast the Uranium has decayed as you don't know all the conditions and environment it's been in. This is the unknown regarding the 'Rate of Decay' and the BIG hole in dating accurately. We are not talking just a few years, there could be millions of years difference.

There have been some interesting examples where rocks of a known age have been sent for testing. One from a recent volcanic eruption was dated at 3.5 million years when it was only 50 years old.

Another well-known one was from the 1980 Mount St. Helens eruption. I remember it well as my wife and I were flying from Vancouver to Miami just after the top of the mountain blew off. The flight path was close to the caldera. I will never forget the magnitude of the plume of smoke rising up, an awesome sight of the unbridled power of nature.

After the mountain blew its top, newly formed lava continued to flow for a few years, creating new rock. Once it stopped flowing rocks formed and a chunk of that rock was sent for Radiometric dating. As it had formed from new lava flow, they knew the rock was only 30 years old, no questions, no assumptions.

This sample was tested using Potassium(parent)–Argon (daughter) dating.

The Radiometric dating showed the rock was 350,000 years old and the minerals were 2.4 million years old. It was unquestionably a 30-year-old rock, need I say more. This shows undeniable and irrefutable evidence that the dating methods and decay rate assumptions used by evolution scientists are seriously flawed. It's a con.

Undeniable Scientific Evidence Evolution is a Con

You have discovered compelling facts proving The Big Bang and evolution has not been scientifically proven and is instead merely a concoction of assumptions and theories.

If an evolution scientist's imagination conceptualises an idea or theory and their like-minded cronies agree it sounds plausible, then in their science fiction world it becomes established truth. They admit this themselves. Professor Richard Dawkins, a leading atheist and evolutionist wrote; *"Science proceeds on intuitive leaps of the imagination building an idea of what might be true"*.

Here is another classic admission from Dawkins and Christopher Hitchens, another outspoken atheist and evolutionist; *"we have a working theory we know is true"*.

These quotes provide further evidence how deluded their minds have become from the endless theories they spin.

They have lost sight of what the real meaning of truth is. As we know, based on age-old and long-stablished scientific principles, a theory is an idea, notion, speculation, supposition or view, it is not truth until it is proven.

Hopefully, the mainstream scientific community doesn't share this mindset or we are in big trouble. Scientist's findings would no longer be trusted as we would have no way of knowing if they are based on true science, tested and proven, or is just a theory based on their 'imagination' and 'assumptions'.

In this chapter you will discover undeniable biological scientific evidence that evolution is impossible. You will uncover how this evidence comes from one of the most common species on earth. It is clear, concise, scientific fact and we can see it every day.

It's a fact evolutionists can't deny, nor can they provide scientific evidence refuting it if they are willing to be honest.

This chapter creates a real Big Bang; it completely blows up the theory of evolution.

Undeniable evidence evolution is a con

Nature reveals a level of incredible design and detail that is so meticulous it defies common sense reasoning how such minuscule and intricate detail could have just randomly evolved over billions of years without design and in a perfectly ordered process.

If just one tiny part in the process didn't perfectly align with the others it would not have supported life, preventing the animal from forming. Scientifically the odds of this happening is impossible.

The incredible design of animals and nature has been an interest of mine for many years. Their complexity and beauty never fails to amaze me, and I could give thousands of examples that would astonish you. I could write another book just on that topic, maybe I will one day.

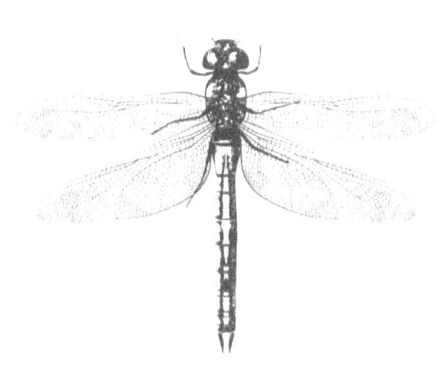

Here is one specie worth a quick mention, a little fly that gives evolutionists a horrific headache.

The Dragonfly can hover, fly backwards and manoeuvre in tight spaces. It was the model used by Igor Sikorsky, the Russian scientist who designed and produced the helicopter. That's how fantastic the design in nature is, it just doesn't happen by random chance.

The Dragonfly's eye has 30,000 lenses and takes in 200 images per second. These eyes are more sophisticated than the best HD camera and apparently, they can see in colour, ultraviolet and polarized light.

It would be impossible for such an amazing design to have simply evolved, there had to be an intelligent designer behind it. Evolutionary scientists tell us they have found a dragonfly fossil 300 million years old preserved in amber. What's worth noting is this primordial dragonfly looks EXACTLY like those of today. This amazing insect hasn't changed in the past alleged 300 million years.

As you read in chapter four, despite their claim, it's not 300 million years old. However, if we stay with their story isn't it interesting that it's still the same, no evolution.

The explanation evolutionists give is the dragonfly is an 'anti-evolutionary specie'. This is a new term they invented when they found a species that hasn't evolved, contrary to their theory that all species evolve into a better specie. This means for some unexplained reason evolution stopped working in this one specie. It somehow decided the dragonfly was perfect and didn't need to evolve over 300 million years, try to figure out that one!

You never hear the dragonfly mentioned by evolutionists as it provides evidence evolution is a con.

When you study the complexity of nature, you discover how everything had to be in place at a single starting point for it to survive. If insects weren't perfectly formed as we see them today, they couldn't survive.

Once we understand the complexity in their design, we see just how absurd the theory of evolution is. The simplicity of nature is the strongest test on whether evolution could be true.

You are about to discover a common specie as proof it had to be created. Its complexity will amaze you and prove categorically nature is far too complex to have evolved on its own.

What you are about to read is biological science, it's proven, verifiable, compelling, uncomplicated and true, proving without a doubt evolution is impossible.

First, a question for you to ponder; **what came first, 'The Chicken or the Egg'?**

An age-old question upon which much debate has taken place.

Before reading on, please STOP... think about it, then write down your answer and reason.

This question inherently has some challenges in answering it with certainty, which is why it has fostered much debate over the centuries.

If you answered the 'egg'; that's wrong as there needs to be a chicken to lay an egg, no chicken, no egg and no breakfast.

If you answered the chicken; how could that be true as it takes an egg to hatch into a chicken to lay the egg, no egg, no chicken?

A huge problem for evolutionists, isn't it interesting that one of the oldest and most basic questions would turn out to be the downfall of scientists and evolutionists?

Typically, they have plenty of theories based on their imaginations and assumptions which, in their science fiction world, they believe are true. They are forced to come up with something as they claim evolution is true and supported by science, the science of their imagination,

I won't bore you with all the various theories from their imaginations, some are just so crazy you would think I'm making it up.

The most common theory put forward is... the egg came first, but not the egg as we know it today. Prepare yourself, here is an example of the wonders of an evolutionist's imagination explained by a college professor.

***There is no such thing as a chicken theory**–Chickens aren't a static thing, they evolve over time, they are constantly changing and so are their eggs. Millions of years ago there was a dinosaur that looked vaguely like a chicken, but it had teeth and claws on its wings. If you saw it at night, you might briefly mistake it for a chicken.*

Wow, it's a shame they evolved. Imagine the size of their eggs, one of them would feed an entire family for breakfast!!! Hold on it gets better.

Over time, though, this creature changed. Its teeth disappeared, as did the claws on its wings. It gained the ability to fly and then lost it again.

At what point did it become a chicken? It still isn't a chicken, remember? There is no such thing as a chicken.

The eggs you buy at the store come from a small dinosaur still in the process of becoming what it will eventually become. It is the first of its kind. It is the last of its kind. Its children will not be chickens, any more than it is.

I haven't included the professor's name as I don't want to embarrass the poor man.

And that was one of the saner ones... According to scientists, all the eggs you have been eating over the years are really dinosaur eggs!!! I bet that's a big shock to you. So why are they called chicken eggs if we should call them dinosaur eggs? He goes on with even more unfactual rants from his "imagination".

There is absolutely no science behind this. Typically, when faced with a question for which they haven't and will never have a true scientific factual and proven answer, we get a serving of their science fiction nonsense!

Remember in chapter 1 where we listed the criteria for a scientific discovery. There are NO SCIENTIFIC FACTS to substantiate his story. And this is a college science professor, what other rubbish are these guys teaching.

If you believe scientists, there was no first chicken or egg, and you are eating dinosaur meat and eggs!!! KFC has been using an incorrect brand, they should be named KFD, 'Kentucky Fried Dinosaur', yum that sounds more appetizing.

Here is another answer from a scientist's imagination; when cornered and he had to give an answer this is what he said:

But it doesn't matter; at some point in evolutionary history when there were no chickens, two birds that were almost-but-not-quite chickens mated and laid an egg that hatched into the first chicken. If you are prepared to call that egg a chicken's egg, then the egg came first. Otherwise, the chicken came first, and the first chicken's egg had to wait until the first chicken laid it.

Another typical example of their nonsense and the gaping holes in evolution. They have no intelligent answer how the chicken or egg evolved with all its complexity, as you are about to discover. Do these guys ever read what they say to learn how stupid they sound.

Leaving the chicken out, they claim the egg, as we recognize it today, first came on the scene with the evolution of a species that birthed a fetus in a membrane millions of years ago.

Prior to this, they claim most animals relied on water reproduction, laying their eggs in ponds so they wouldn't dry out. Then at some unknown point in time the egg, all on its own, decided it needed a shell, so it decided to evolve, designing an egg structure. They somehow worked out what to make it from, and how their bodies could produce the required minerals in just the right proportions and shape to fit within their own bodies. All this was done by a simple undeveloped life form with a brain the size of a pinhead, yeah sure!!!

They admit they don't know when this happened as eggs don't make good fossils, so they have no evidence to support their claims. They have no evidence, just *'our best guess'*, the delusional imagination at work again

I'm not a scientist as you no doubt have gathered, I'm just a common-sense guy wanting to know what is true and does research to find out the answer. Even I can come up with a major problem that blows away their feeble explanations. This is a 'clanger', that's my terminology for an awesome statement of fact that blows evolution scientists and their theory of evolution away.

Chickens form eggshells from a special mix of calcium carbonate and **Ovocleidin-17, which is only produced** inside the mother hen. Without a fully formed and functioning mother hen possessing a fully formed and developed internal egg production system, there would be no production of **Ovocleidin-17** to make the eggshell. This is undeniable scientific evidence and proof evolution is impossible. The chicken was created, no other alternative.

Here's another one, the eggshell MUST be fully formed, including **Ovocleidin-17, inside** the very first chicken as the shell must protect the embryo of the baby chicken.

You will read further on in this chapter that the formation and specific shape and structure of the egg needs to be precise for a chicken to hatch. It is impossible for an egg to evolve; it has to be perfect from day one or nothing will hatch. It is impossible for an egg to evolve; no eggs would hatch unless a mother hen is fully formed first. Evolutionists claim a specie evolves very slowly from millions of random mutations over millions of years, but with this model no eggs would have ever hatched so the chicken would have become extinct.

Therefore, the judge and jury's verdict is... a fully formed and developed chicken had to come first.

I love it when something as simple and humble as a chicken and an egg confounds what are supposedly the smartest minds in the world. Whilst they throw around many theories and make endless assumptions, they fail to look into the complexity of the little chicken and egg to see how stupid and impossible their theories are.

To further illustrate how little scientists know about their theories, they can't agree if birds evolved from 'Archosauria' dinosaurs, which include cold blooded alligators, or 'theropods' which are the flesh-eating variety; these being enormous creatures so they would have to shrink considerably to become a bird. Scientists who love science fiction stories probably got the idea from one of their favorite movies, *Honey I Shrunk the Kids*. Besides having to undergo a major size reduction, they had to toss away two legs, turning them into wings. Or perhaps it was the two-legged variety of dinosaurs, who knows... they don't.

Ponder this. Why would a dinosaur, the king of the jungle at the top of the food chain want to change itself into a tiny defenseless chicken? This also bring up another question. If just some theropod dinosaurs evolved and others didn't, as soon as the ones that evolved into chickens reached the stage where they couldn't defend themselves, they would be eaten by the 'theropods', making them extinct before they could become a chicken.

I can come up with so many more reasons why this is all so illogical.

However, putting that aside, the bigger issue scientists don't address is dinosaurs are reptiles and cold-blooded, having many biological differences than birds. In fact, **TOTALLY** different, not related in any way. We aren't talking a few changes here like an arm turning into a wing, growing feathers, or other simple stuff like that!

Birds have very high body temperatures. The chicken, for instance, is around 106 Fahrenheit resulting from a high metabolic rate. As you will read further on, if it didn't have this higher body temperature, a baby chick could never hatch from an egg and the species would be extinct even before it evolved. If the evolutionist claimed the dinosaur/chicken increased its body temperature over time, this couldn't happen. Before it became a fully developed chicken its body temperature would never be high enough to form an egg properly and no eggs could hatch. Or perhaps one day the dinosaur decided to be a chicken. With its superior intellect it knew that to fulfil its lifelong dream to become a chicken it needed a warm climate, so it packed its belongings and moved to Florida.

The other issue is maintaining a constant body temperature. Birds and mammals, including humans, have mechanisms to maintain their body temperature. However, reptiles' (dinosaurs) body temperatures change according to the surrounding environment.

There are also many other significant differences between dinosaurs and birds, meaning the old dinosaur would have needed a complete redesign and remodeling. To even suggest a chicken came from a dinosaur is so ridiculous it's a joke, sorry scientists but really.

To top it off, there are no fossil remains of dinosaurs with feathers, all they have is skin or scales. I guess the evolutionists will try to tell you feathers evolved from scales. Once again, a huge difference as feathers grow from hair follicles while scales are a skin on a body. The old dinosaur would have to go into the body shop for a total rebuild.

Do evolution scientists know how stupid this stuff is? They are trying to tell us that one day a dinosaur decided it wanted to be a bird; so it told its children they should also be a bird and if they really concentrated and believed in the power of positive thinking they will change slightly, just ever so slightly, so slightly they won't notice the change. They won't actually become a bird themselves as this will take millions of years to happen, so your great grandkids, a million generations down the family tree will one day be the finished product. It will take a long time, however it will definitely be worthwhile as we will go from being king of the beasts to a tiny little chicken. Then the real benefit will come as whoever takes our place as king of the beasts will end up eating us for dinner. That will be so great for our offspring, so make sure you tell your children, who tell their children, who tell their children for a million generations. Then, if everyone in the family line remembers to tell their children, and if each generation uses the power of positive thinking to change some part of their body, it may work, and we will reach our ultimate destiny. We will become a tiny little defenseless chicken. Nothing against chickens, of course.

Goodness knows how all the generations came to agree on which parts needed to be changed first, a mere technicality in the minds of the evolutionists.

Hello, what dinosaur in its right mind would want to change from the king of the jungle, top of the food chain to a little old timid chicken that gets eaten.

Evolutionists will tell you they had no choice. It was the mysterious force of evolution deciding the earth couldn't sustain dinosaurs any longer, so they had to became extinct. That's also rubbish and the topic of another book.

It's a science fiction fantasy story, there is no evidence of any transmutation specie from dinosaur to chicken. NOT ONE example. No evidence, just another stupid theory which holds as much weight as my theory of superior intelligent Martians from another galaxy located millions of light years away in a black hole leaving behind a group of Martian workers who built the earth and animal kingdom. Don't know why we can't see them, but don't worry, I'll come up with a theory from my imagination to explain it away. Who knows, one day it may even end up in school textbooks, another unproven theory taught to children!

Real proven science has also proven evolutionists wrong. The Ovocleidin-17 I mentioned earlier was discovered in July 2010 by British scientists using a supercomputer. This led them to claim they finally came up with the final and definitive answer to which came first, the chicken or the egg.

They identified the protein, Ovocleidin-17 speeds up the production of eggshell within the chicken and is a vital component as the egg inside the mother chicken must be ready to lay within 24 hours. Ovocleidin-17 is **only produced** by a chicken, so a chicken egg can only come from mother chicken, the hen. So, science proves the chicken had to come first in its current chicken state; it had to be created, as there needed to be a fully evolved chicken to produce the Ovocleidin-17 needed to form an eggshell.

They also proved it would be impossible for mother hen to have evolved, as a complete egg with a fully formed shell is made in only 24 hours inside mother hen.

The evolutionists would probably claim the dinosaur that moved to Florida to evolve into the chicken learnt how to produce the protein, Ovocleidin-17. Then one day it decided that if it really wanted to be a chicken, it had to somehow change its body mechanism to produce the special protein, along with all the other changes required to be a productive chicken. Yeah sure, shame on me I'm getting so cynical!

Once again, I want to stress I'm not getting after all scientists, there are some wonderful, very intelligent scientists who are doing great work. Science has come a long way over the decades with many very smart scientists developing very useful things, especially in the area of medical research, which saves lives and makes the earth a better place. These scientists are great, and we appreciate their discoveries. Keep up the good work.

It might be the productive scientists have evolved further than the evolution and theoretical scientists who are still evolving into more useful scientists!! Wouldn't that be nice.

The ancient Greek philosopher Aristotle, who lived from 384 BC to 322 BC had the answer over 2,000 years ago.

He wrote: *"There could not have been a first egg to give a beginning to birds, or there would have been a first bird which gave a beginning to eggs; for a bird comes from an egg."*

After reading the nonsense from today's academics on what came first compared to Aristotle's view, it provides further evidence humankind is not getting more intelligent.

As I said at the beginning of this chapter, the complexity of nature is the best test of whether evolution could be true. What you are about to read will scientifically prove the designs in nature are just too amazing to have ever evolved on their own. It is some of the most compelling, simple and understandable evidence that evolution is a con.

How the humble chicken and egg conundrum proves evolution theory wrong

When you see eggs on your breakfast plate, do you ever stop and think about the incredible achievements of the humble dinosaur, whoops sorry... Chicken in manufacturing an egg. The complexity is a wonder to behold.

To enable a baby chicken to hatch, incredible design features have been built into the egg. All these MUST be in place to the finest detail or there would be no eggs or chickens. It's so incredible that it proves evolution is impossible.

Everything in the egg has to be perfect and happen in a particular order and on a particular day. If one small thing goes wrong there would be no chickens and no KFC.

The beginning of an egg starts with the yoke. Contrary to what most people think, a yoke forms first in the ovary system of the hen, then as it forms it passes through the oviduct. Most females in a species have two ovaries, but birds are unusual as they only have one. The ovaries and oviduct take up a relatively small area inside the chicken, considering the oviduct when stretched out is around 2 feet long with five distinct sections.

1. Ovary
2. Infundibulum
3. Magnum
4. Isthmus
5. Shell gland
6. Vagina
7. Egg

1. Ovary

It converts food the chicken eats into nutrients, forming the building blocks of the egg yolk, these are:

1/3 Protein

1/3 Fat

1/3 Water

These are carried by the bloodstream from the liver to the ovary. Once in the ovary, tiny tissue bags called follicles fill with the yoke and grow. The largest follicle will release the yoke of the egg the hen will lay tomorrow, while the next largest will produce an egg the next day etc., and so on.

In 1 - 2 weeks a follicle will grow from 1 mm to the mature size of 25 mm. When a yoke matures the follicle ruptures along a line relatively free from blood vessels, the stigma and yoke is then released. If any blood vessels cross the stigma, a tiny spot of blood may spot the yoke as it's released from the follicle. This explains why sometimes when you break an egg open you may see a small spot of blood on the yoke.

2. Infundibulum

The yoke now moves into the funnel-shaped infundibulum at the upper end of the oviduct where it spends about 15 minutes here. This is the only time the yoke isn't covered with albumin, the name for an egg white. Fertilization, if it is to take place, will happen here. Then, over 24 hours, the yoke moves down the oviduct.

3. Magnum

The yoke then passes into the Magnum and gets covered with a dense shock absorbent layer of albumin (egg white) over the course of about 3 hours. As it forms in the magnum, spiral ridges running the length of the magnum cause the yoke to spin. This process twists the fibres in front and behind the yoke, making two pigtail like rope structures called 'chalazae' designed to keep the yoke suspended in the center of the albumin and ultimately prevent the yoke from moving around inside the egg. These are the little white bits you sometimes see in the egg white. They play a vital part in the egg's design and are essential for successful hatching. Without the chalazae **no chick would ever hatch**, there would be no chickens or eggs for breakfast. This is how intricately the egg is designed and made. How could a dinosaur that changed into a chicken know how to do all that, foolish suggestion?

1. **Isthmus**

The Albumin covered yoke then moves into the 'Isthmus', this is the shell manufacturing factory where the first layers of shell membranes are deposited and wrapped loosely around the albumen covering the yoke. During this process the egg looks more like a prune.

2. **Shell gland**

The partially formed eggshell passes into the shell gland where the shell forms over the next 20 hours. First, a thin albumin is secreted which is mostly water moves through the two shell membranes into the highly concentrated thick albumin surrounding the yoke by osmosis. This pumps the egg into a normal shape and structures the shell membrane tight around it. Next, the shell gland secretes a highly concentrated solution of calcium carbonate. This is the Ovocleidin-17, which is only produced by mother hen. Crystals of calcite form and grow on the outer shell membrane. These are like little triangular crystals that grow into each other over the egg, eventually joining each other. Tiny pinhole spaces are left between the crystals, leaving tiny air holes in the shell. These are crucial as without them a baby chick will not live and hatch.

Lastly, a special protein solution called the 'cuticle' is deposited onto the eggshell. Gas can pass through, but the two layers protect the egg from harmful bacteria. If this wasn't perfect from day one bacteria would kill the baby chicks so they could never hatch.

3.   Vagina

Finally, in a process called 'over position' the egg flips end over end. This occurs through contractions of the uterus until it pushes the egg out of the hen's body and you have one perfectly made, incredibly designed egg. And evolutionists claim this all just happened by random chance, there is as much chance of that happening as the government abolishing taxes, or evolutionists admitting they got it all wrong.

4.   Egg

Even after an egg is laid the process isn't finished. As soon as it's laid, the yoke and albumin fills the entire shell, however a hen's body temperature is around 106 Fahrenheit but an egg is typically laid in an environment of around 20–40 Fahrenheit (66-86f). As the egg cools its contents contract, with an air cell forming between its two shell membranes. During the hatching process a chick will puncture the membrane and breathe in air from this cell before hatching. If this air cell didn't form the chick couldn't hatch.

Whilst that's amazing, it gets better when we learn why some of these things take place.

An eggshell isn't watertight, it's covered with about 10,000 very tiny air holes that are so small we can't see them. This explains why when you boil an egg you see small bubbles leaving the shell.

These air holes allow air to enter the membrane so the chick can breathe inside the shell. If a perfect amount of airholes didn't exist, the chick would suffocate. Think about the number of eggs laid in one day across the world, all with a perfect amount of airholes. Another mystery for evolutionists, if things keep evolving why do some things stay at a constant state. If the egg evolved just slightly, and the shell when forming didn't have perfectly formed tiny air holes then chickens would become extinct in a few weeks and humans could not rectify this. It gets even worse; KFC would go out of business.

If you have children concerned the egg-yolks they eat are a future chicken, I have good news for them. When the chick is first formed it is a tiny little embryo fixed to the outside of the yoke, the yoke doesn't become the chick. If the egg isn't fertilized there is no embryo baby chick inside it.

By the fifth day the little chick, safe and warm in its shell, sends out two blood vessels that grow out towards the shell and hook into the shell membrane. This is a very thin skin lining the inside of the shell. You would have noticed it when you peel a boiled egg.

One vessel is used by the chick to breathe and the other is used to get rid of waste. As a chick grows inside the shell it eats the yoke, this is its source of food, accessed from two more vessels coming out of the chick and hooking into the yoke.

If these don't happen in the right way and at the right time, the chick won't survive.

Here's another question, how does mother hen know how to hatch the egg? It must be kept at an exact temperature to hatch. She uses grass, straw or twigs in a pile which will decompose, creating the heat required. How does she monitor that temperature, how does she know if it's too hot or too cold? All this has to be perfect from day one, this couldn't evolve a bit at a time, it had to be exactly right from the start, or no eggs would hatch. It couldn't evolve through trial and error.

Inside the egg some other incredible things are happening.

The chick must be held in a certain position within the egg. If this doesn't happen, when it's time to hatch the chick won't be able to get out of the shell.

Whilst mother hen is incubating the egg she moves it, and if the chick isn't held in exactly the right position it won't hatch. To keep the chick in place it's hooked onto the yoke which must be kept in place. This is done with two little ropes of albumen attached in a mysterious way from either side of the yoke to the inside of the egg membrane. These are the little white bits you see in the egg white when you crack a raw egg open.

The way it attaches these ropes is incredible as no amount of spinning the egg will break them. An egg can spin around and the ropes will stay in place, holding the chick in exactly the right position.

Why does the chick have to stay in position? The egg has a hard end and a soft end. It must hold the chick's head in the soft end where a little air sac is positioned.

You would have noticed when you boil an egg and remove the shell the egg doesn't fill the entire shell; one end is recessed. This is where the air sac is.

Exactly on the 19th day the little chick has to break open the shell.

His beak is very soft and not capable of cracking open the hard shell, so it has a special shell breaking tool. A tiny cone made of a hard substance perfectly fits over the bill of the baby chick, enabling it to break the eggshell.

If it didn't have this cone it would never hatch. It couldn't evolve, it had to be in place on the first chick in the first egg.

To give the chick the extra strength needed to break out of the shell, the little air sac at the top of the egg has just enough air to last for 2 days. Therefore, the chick must be in exactly the right position to reach the air sac. If it didn't have this air, it would suffocate.

As air is depleted the chick knows it must break out, so it takes a big lunge at the remaining air in the sac, and using all its strength to get more air its beak cracks through the eggshell.

The chick then struggles out of the shell. Even the action exhibited by the chick to break free is important and you mustn't help it as its struggle to get free helps strengthen its wings and lungs. This action also disconnects the blood vessels connected to the shell. If you help, the chick may not develop properly and could die.

Two days later the little nose cone falls off its beak.

The little chick instantly knows its mother although it has never seen her.

For scientists and evolutionists to claim the egg evolved over time is absolute rubbish, how could it. If just one of those design attributes is missing, eggs don't hatch. It had to be created as it is today for the first chicken to hatch. The chicken had to be designed and built at one time. It simply had to be created, there is no other sensible explanation.

Now that's something to ponder.

If you're an evolutionist, and despite this overwhelming, undeniable evidence you still blindly believe in evolution, let's explore your theory.

The chicken evolved from a dinosaur who became a bird who became a chicken. If that was true, then we are dealing with a superior highly intelligent lifeform with the intellectual power to plan out the egg's design and instantly build the egg-making internal organs within their bodies.

We know without a doubt the egg couldn't evolve, so during one of those evolutionary stages one of them had to design and build the internal egg manufacturing system overnight so it could produce an egg.

They only had one chance to get it right, if they stuffed up and got just one small part of the design wrong, no eggs would hatch. They would eventually die, and the specie would become extinct.

For your dinosaur/bird/chicken evolution theory to work, one of these had to be a super intelligent lifeform, possessing an intellect far more advanced than mankind. Even mankind using super computers and artificial intelligence can't design and build a lifeform as impressive as a chicken.

Or, as evolutionists have such a great imagination perhaps, we could look at another possibility.

One day some chickens that evolved from dinosaurs got together for a think-tank session once they discovered they couldn't produce eggs and would very quickly become extinct. Combining their design ideas, they picked one lucky chicken to be the test chicken, the prototype. Then, their top surgeon operated on the prototype chicken, installing newly designed egg-producing organs assembled from old dinosaur parts. Miraculously, he got it right the first time, and it produced the first egg in the prototype mother chicken, saving the specie.

An amusing and silly analogy, however in principle this is what evolutionists claim.

An incredible feat, but nothing is impossible in the imaginations of evolutionists. Somehow, chickens or dinosaurs, or whatever, knew the design and how to build the little beak cap so the chick could break free.

It then designed the shell with 10,000 little air holes of just the right size, too small not enough air, too large the egg white would escape.

And the cord, just thinking of it in the first place and the incredible design, the composition. How it's held in place on the yoke and shell membranes, how it holds the yoke perfectly in place so the chick can reach the air sack vital to hatching.

How they designed and built the chick's brain and pre-programed it so it would know what to do inside the egg and hatch on just the right day, remembering chickens hatch at the same time, 21 days.

Also, the invention of Ovocleidin-17, which no human or specie has replicated or invented. It's unique to the chicken.

Even the chicken's internal eggshell manufacturing mechanism produces the exact type of calcium carbonate so a shell would form inside the chicken. Otherwise she would lay scrambled eggs and mankind would have never enjoyed a poached or boiled egg.

What incredible intelligence those little chickens have to design and build all those intricate, highly complex components.

I could continue, but I'm sure you're getting the picture; one needs a massive imagination and amount of faith to believe all this just evolved.

According to evolutionists, the dinosaur chicken would have to be a highly super intelligent lifeform far superior to mankind. Hopefully chickens never see a *Planet of the Apes* movie. It might give them the idea to rebel and take over the world. We would be outnumbered, as I read there are around 25 billion chickens in the world, and if we didn't eat eggs there could be trillions in a few weeks.

Without doubt what I have written about is a million times more insightful than Darwin's ramblings, and provides clear evidence of how impossible evolution is. I somehow doubt evolutionists will give me the same accolades they gave Darwin, who proved nothing.

They may give me the title of the person who destroyed their evolution con and saved them from all the time they spend on their rubbish theories so they can convert to a productive field of science and be an asset to humanity.

Looks like a very convincing argument, providing simple proof the theory of evolution is flawed and a con.

There is another big problem for evolutionists that credible scientists have discovered, proving yet again evolution is impossible. Chapter six makes fascinating reading.

WHY DID THE EVOLUTIONIST CROSS THE ROAD?

To re-experience "life at the chicken stage."

Science of DNA proves evolution impossible

In the 1800s proponents of evolution theories didn't have the insight into the amazing world of true science we have today. Not theoretical nonsense used by evolutionists, but real proven sciences such as how a human body functions, anatomy, physiology, histology, embryology and the amazing recent discoveries of DNA and genes.

True science supported by reliable established scientific methods is also proving evolution impossible through the incredible complexity of the DNA code.

Whilst evolutionists make many false claims that science has proven evolution true, in reality it's actually the reverse. Modern discoveries are proving the impossibility of evolution by revealing its many flaws, along with scientists' theories and assumptions. The evolution scientist's credibility is crumbling.

Among these discoveries are the incredible developments in our knowledge of the building blocks of life, DNA, Proteins, RNA, Enzymes, Genomes and Chromosomes.

In this chapter you will discover this amazing complex world and the wonders of DNA and genetics.

Darwin's writings were based on his opinions and theories as to how species evolved over time. He had NO HISTORICAL PROOF for his theories, as in his lifetime nothing would have changed.

Darwin's observations and assumptions were primitive, as he had no knowledge of DNA. The field of genetics did not exist until the 1900s.

If Darwin had known about DNA, his books would have been very different. DNA is a game changer, proving his assumptions wrong to the point his books are redundant.

Modern science has made amazing discoveries, not theories but actual proven discoveries, relegating Darwin's evolution theories to the junk heap.

There is even a 'Knock Knock joke reflecting how primitive his findings were.

"Knock Knock."

"WHO'S THERE?"

"Darwin."

"DARWIN WHO?"

"That's what they'll be saying 50 years from now - Darwin WHO?"

In the past 50 years, discoveries in DNA have opened a whole new window into how species, including humans, are built. You have probably heard on the news how DNA is solving murder cases, some going back many years as it holds a unique genetic code for each person.

DNA is very complex, and as you about to discover, any rational common-sense person will conclude DNA could never have evolved.

As an analogy of what DNA evolving relative to its complexity would be like, imagine putting billions of numbers and sequences into a big balloon. Then fill it with air so it explodes, causing all the numbers to fall on the ground, forming a code more sophisticated than anything the most intelligent human brain or super-computer could create. That's how ridiculous it is to even suggest DNA evolved.

DNA code is so advanced and intricate that frankly, if anyone thinks it happened just by random chance, they need a serious reality check.

Genealogists use DNA to verify if two individuals are closely related, as a person's DNA gets passed on to the next generation in big chunks called chromosomes. Every generation, each parent passes along half their chromosomes to their child. Most people have 23 pairs of chromosomes for a total of 46. One of each pair comes from mom and the other from dad. Therefore, we are 50% related to both our parents.

You may have read about scientists experimenting with gene editing to eliminate inherited diseases.

In 2018, a Chinese scientist claimed to have created the world's first genetically edited baby while trying to insert the ability to resist HIV infection. (*The Guardian Nov 26, 2018 – Gene editing*).

This Chinese scientist, He Jiankui, sought to disable a gene called CCR5 that forms a protein doorway allowing HIV, the virus that causes AIDS, to enter a cell. He did the gene editing during in-vitro fertilization when the gene-editing tool was added. When the embryos were three to five days old, he checked them for editing. Errors could happen in this process, which is why his process is surrounded in so much controversy, as they don't know what they are experimenting with and the potential flow-on effects to subsequent generations.

There has also been much dialogue on using DNA editing to make designer babies.

Medical Scientists discovered how to edit genes, the strands of DNA that govern the body. The tool, called 'CRISPR-Cas9' technology, makes it possible to make precise changes in any cell or organism. (CRISPR stands for 'Clustered Regularly Interspaced Short Palindromic Repeats' - Cas9 is short for 'CRISPR associated protein 9')

Editing DNA can supply a needed gene or disable a gene causing problems. For example, it can disable a hereditary disease like Cystic Fibrosis, possibly preventing it from being passed to the next generation

It works this way; The CRISPR/Cas9 system occurs naturally in bacteria and gets its DNA-cutting abilities from its role as part of the bacterial immune system. Snippets of DNA from invading viruses are cut and stored in the bacterial genome as part of the

CRISPR array.

The Cas9 protein uses those snippets to recognize future invaders and cuts their genetic material, killing them. The CRISPR/Cas9 array allows the bacteria to recognize future attacks and, because it becomes part of the bacterial genome, it passes that immunity on to its offspring.

There is a video on *The Guardian* from Professor Jennifer Doudna, one of the pioneers of 'Crispr-Cas9' gene editing. She describes how, early in the technology's development one night she had a dream where a scientist introduced her to a man in a dark room. When the man turned around it was Adolf Hitler, asking her to describe how the Crispr technology worked and how it could be useful to him. She woke up from that dream with a real start, motivating her to publicly discuss the implications of this new technology. Wow, what a powerful dream, especially considering what you read in chapter two.

She also answers the interesting question, will they be able to make a human? She replied, *"in reality it will not be possible to design a human being, we are too complicated and there are too many unknowns about the human genome"*.

Even the best minds admit they could never create a human. We are so complicated they don't understand how we are made, yet evolutionists try to tell us the human body just evolved from millions of random mutations with no designer, it all just fell together.

The vast progress made in understanding DNA has also shaken evolutionists and discredited their theories yet again. Naturally, they won't admit there is a gaping hole in their evolution theory.

DNA is a complex chemical module found in all living organisms. It was first discovered in the 1950s. DNA is short for *'deoxyribonucleic acid'*. It's the master molecule of every cell and contains vital information that's passed to each successive generation. It coordinates the making of itself and other molecules (proteins). If even slightly changed, serious consequences may result. If it is destroyed beyond repair, the cell dies.

Important point: Note how DNA coordinates the making of itself. DNA needs to be present to make DNA, it could not have evolved, it is impossible, DNA had to be created at the point of a species' creation. Evolution is IMPOSSIBLE.

DNA is like an assembly manual of organisms, in other words it's the storehouse of genetic information in a living organism, including you, me, the cat and dog. DNA is in each cell of an organism; it stores a huge amount of information and tells the cells which proteins to make.

The DNA in a cell is a pattern made up of four different parts called nucleotides. A nucleotide is one of the structural components or building blocks of DNA and RNA. A nucleotide consists of a base (one of four chemicals: adenine, thymine, guanine, and cytosine) plus a molecule of sugar and another molecule of phosphoric acid.

Imagine a set of blocks with only four shapes, or an alphabet with only four letters. DNA is a long string of these blocks or letters. Each nucleotide comprises a sugar (deoxyribose) bound on one side to a phosphate group and bound on the other side to a nitrogenous base.

Genes are made of DNA. DNA is a long molecule. For example, a typical bacterium like E.coli has one DNA molecule with about 3,000 genes (A gene is a specific sequence of DNA nucleotides that codes for a protein.).

Complex organisms like plants and animals have between 50,000 to 100,000 genes on many different chromosomes (humans have 46 chromosomes). Within the cells of these organisms, the DNA is twisted around bead-like proteins called histones. The histones are also coiled tightly to form chromosomes, which are located in the nucleus of the cell. When a cell reproduces, the chromosomes (DNA) get copied and distributed to each offspring, or daughter cell.

The discovery of DNA was a big deal and probably the most important discovery of the last century. Its effect on scientific and medical progress has been huge, allowing identification of the specific genes causing major diseases, which aids in the development of drugs to treat them.

Although DNA shows scientists the building blocks of living organisms, it's tiny. If unwound, the tiny dot of coded information is about 2 meters long.

DNA has the genetic alphabet for a cell to make **proteins**. Proteins are the molecules our cells need to function properly, it's like food.

Proteins have many functions:

- Enzymes that carry out chemical reactions (such as digestive enzymes)
- Structural proteins that are building materials (such as collagen and nail keratin)

- Transport proteins that carry substances (such as oxygen-carrying haemoglobin in blood)
- Contraction proteins that cause muscles to compress (such as actin and myosin)
- Storage proteins that hold on to substances (such as albumin in egg whites and iron-storing ferritin in your spleen)
- Hormones - chemical messengers between cells (including insulin, oestrogen, testosterone, cortisol, et cetera)
- Protective proteins - antibodies of the immune system, clotting proteins in blood
- Toxins - poisonous substances (such as bee venom and snake venom)

The particular sequence of amino acids in the chain is what makes one protein different from another. This sequence is encoded in the DNA where one gene encodes one protein.

DNA and protein are the two basic components of all lifeforms. Even a virus has DNA and protein. Scientists are discovering how they can modify a virus's DNA so they self-destruct. Now this is worthwhile, wonderful scientific work which will help humans.

If you want to discover more on the amazing structure of DNA, google 'how proteins are built'. Their complexity is amazing; there is no way it could just evolve by random mutations over billions of years. I have read there could be something like 10 trillion cells in a human. This means each person has around 60 trillion feet or around 10 billion miles of DNA inside them. Putting that into perspective, Pluto is "only" about 3.67 billion miles from the sun.

(Dr. Barry Starr, Tech Museum & Stanford University FEB. 2, 2009)

So, evolution scientists are trying to tell us that each person's highly intricate DNA, all that 10 billion miles inside of us with all that complex coding, just randomly fell into place.

No intelligent design, just by mere chance from random mutations. Do you see how ridiculous evolution is and why developments in real science are constantly proving evolution is redundant?

Both **DNA** and **proteins** are long molecules **made** from strings of shorter building blocks.

While **DNA** is **made** of nucleotides, **proteins** are long chains of amino acids that form the basis of life and consist of a group of 20 different chemicals, alanine, arginine, asparagine, aspartic acid, cysteine, glutamic acid, glutamine, glycine, histidine, isoleucine, leucine, lysine, methionine, phenylalanine, proline, serine, threonine, tryptophan, tyrosine, and valine.

For protein to form a cell it needs to put a chain of amino acids together in just the right order. **It can't evolve over time, it has to be perfectly formed the first time around**; the protein is an engineered system with a specific amino acid sequence

Here is another evolution killer. It starts by making a copy of the relevant DNA instruction in the cell nucleus and takes it into the cytoplasm, it copies what's there already. If that was still evolving, what it copied would not work as it's not in a complete state and ready to function as the cell decodes the DNA instruction and makes many copies of the protein, which are folded into shape as it produces them.

Please tell me how evolution is possible in this scenario. It's time to bury evolution.

It should be laid to rest under a tombstone titled;

"Here lies the biggest con inflicted on humanity"

Amino acids are organic molecules comprising carbon, hydrogen, oxygen, nitrogen, and sometimes Sulphur. It's the amino acids that produce proteins and other important compounds in the human body such as creatine, peptide hormones, and some neurotransmitters.

The 20 amino acids can be arranged in millions of different ways to create millions of different proteins, each with a specific function in the body. The structures differ according to the sequence in which the amino acids combine. Amino acids are linked together by a chemical bond called a 'peptide bond'. As these chains are formed, they are called polypeptides. When one of these polypeptide chains form into a specific shape it can then fit into a cell and form a task, where it then becomes a protein. For this to happen the amino acids must come together in a certain way to make a very specific and unique shape and are arranged in certain ways. **This cannot possibly have happened by random chance.**

Proteins are an essential ingredient in the structure and function of our bodies. We depend on proteins to regulate the body's cells, tissues, and organs. **It takes certain proteins to make the same proteins. This creates a major brain-numbing headache for evolutionists as proteins couldn't evolve into being a protein as you need a specific protein to FIRST form the same type of protein. A specific protein couldn't simply appear on the scene and go into production. It's impossible, so here is another simple point that shows how evolution is a con.**

IMPORTANT POINT - Proteins are like assembly line machines that make all living things, whether viruses, bacteria, butterflies, jellyfish, plants, or humans' function. Each cell in our body has thousands of different proteins, and together these cause each cell to do its job. The proteins are like tiny machines inside the cell.

A human body is comprised of around 100 trillion cells. That's a BIG number, which shows how complex our bodies are. The impossibility of this mind-blowing awesome design forming by random chance over millions of years from just random uncontrolled mutations is simply a colossal lie perpetrated by atheist evolutionists to further their atheistic agenda.

The Human body is a complex and intricate design.

Now it gets even more interesting. Fasten your seat belt, take a breath and hang in there.

There is also another component called RNA which is also needed in the DNA and protein structure and function.

RNA is ribonucleic acid, a nucleic acid present in all living cells. Its principal role is to act as a messenger, carrying the DNA's instructions for controlling the synthesis of proteins. It carries the codes for amino acids that make proteins, although in some virus's RNA rather than DNA carries the genetic information.

How all the DNA, RNA and Proteins interact is **another brain-numbing problem for evolutionists** as, when one tries to get one's head around the complexity of the subject, it looks like they are interdependent of each other. DNA needs to convert into RNA to produce proteins, however proteins are required for DNA to duplicate, and if even one of these VERY complex structures falls apart, so do we.

So, how could DNA, RNA and Proteins evolve over millions of years? They all need each other to survive and would need to appear at exactly the same time. They all had to appear in a specie at the same time and in the same state. Here is yet another example as to the impossibility of evolution. A human had to be created.

Here is another brain-numbing problem for evolutionists.

Think back to the Big Bang and add this into their theory. They claim the gases that miraculously joined together from a big explosion all by themselves formed living cells and then joined with more cells. Then these cells, with no brains, just blobs of jelly-like minuscule things, formed the all the wonders of life and nature we see today. This is what they call chemical evolution.

In DNA there is no attracting force to order the bonding to the backbone of the molecules, no chemical bonding happens, so chemical evolution is impossible.

Evolutionists can't answer this with facts and scientific evidence showing how it happened as there is no answer. They just come up with theoretical rubbish from their overactive science-fiction imaginations.

They can't answer the question of when and how DNA evolved. How could the first bits form when everything needs to be in place from day one for DNA/Proteins to function? Who invented and wrote DNA code? What state was it in in early specie? How did the first undeveloped bit function in the first specie, remembering what you just read on proteins?

As science makes new discoveries on the wonders of life, DNA and Proteins prove scientifically and without a doubt how the beginning of life in their theory of evolution is impossible.

As we have just read, this scientifically puts another nail in the evolutionist's coffin as DNA is very complex and contains a code so complex there is no way it could have just evolved. It's like a super sophisticated computer code.

On July 30, 2016, Bill Gates of Microsoft fame, a man who knows a lot about computer code posted on Twitter:

"DNA is like a computer program but far, far more advanced than any software we've ever created."

Bill Gates is saying DNA is more advanced than anything a human could develop, yet evolutionists tell us it just evolved from lumps of jelly, simple cells with no intelligence developed this complex DNA code. Yeah sure!!! Who are they trying to kid? You've got to go to the meat shop to get more baloney than that.

Science proves DNA is so complex and holds so much information it is impossible for it to have just evolved. It had to be intelligently designed, this stuff is so elaborate and holds so much information that science is still discovering its many wonders.

Each human cell contains approx. 3 billion base pairs* per cell. The DMA DNA of a single cell contains so much information if it were represented in printed words simply listing the first letter of each base would require over 1.5 million pages of text. If laid end to end the DNA in a human single cell measures 3 1/3 feet or 1 meter.

* any of the pairs of the hydrogen-bonded purine and pyrimidine bases that form the links between the sugar-phosphate backbones of nucleic acid molecules: the pairs are adenine and thymine in DNA, adenine and uracil in RNA, and guanine and cytosine in both DNA and RNA. *(Source: The endowment for human development web site)*

Atheist Evolution Scientists are so desperate to prove there wasn't a Creator and evolution is true that they devise sneaky deceptive experiments that, on the surface can look convincing. Here is a classic.

In 1953, two scientists, Miller and Urey orchestrated an experiment in a desperate effort to explain what happened after the Big Bang.

They figured if they could replicate the conditions back then and prove organic molecules could form, then bingo, they would have proof of the Big Bang.

To conduct their experiment, they constructed a sealed system containing hot water and a concoction of the gases they ASSUMED were present 13.7 billion years ago. The mix was their guess as to what was present back on day one, as no one knows. To arrive at their formula, it appears they worked it out backwards. They knew what they wanted to produce, so they created a mixture to produce that result, (H_2O, NH_4, CH_4 and N_2). Then they added an electrical charge to replicate lightning as the catalyst for the chemical reaction to happen. How do they know an energy source like lightning was present before the big bang? I thought they claimed there was nothing or it was some type of nuclear fusion!

They then let the machine run for a few days and afterwards they found various types of amino acids, sugars, lipids and other organic molecules had formed.

They claimed their experiment proved some of the building blocks for life could form spontaneously from simple compounds.

However, DNA and proteins, the code of life didn't appear.

The media jumped in and falsely reported the pair had created life in a test tube. Totally wrong and an example of how this nonsense gets out there where it misleads the public.

What they didn't explain, as previously mentioned, is they really didn't know what gasses or mix would have been present back then, and they purposely left out oxygen as they knew this would kill the molecules they were trying to create.

In other words, they stacked the experiment to arrive at the result they wanted.

The assumption of lightning as a source is also highly suspect. Lightning is an electrical discharge caused by imbalances between storm clouds and the ground, or within the clouds themselves. Most lightning occurs within clouds. What evidence do they have clouds and lightening were present?

Lightning is super powerful, its charge can range from 10 to100 million volts depending on the circumstances, a direct hit would kill and nuke anything.

My experience of lightning would tell me it's not a spark, it's a high voltage bolt that can fry you and would certainly nuke some gases. Once I was walking along a pathway next to the beach while some storm clouds were overhead, and thunder was booming. I knew whenever there is thunder, lightning is nearby. As I was looking out to the sea, a massive bolt of lightning came down and hit the water directly in front of me, around 10 meters away.

This was an amazing yet frightening sight. When the lightning hit the water, it condensed into a ball of fire so intense I could feel the heat it generated. This all only lasted a few seconds but it placed an embedded picture in my brain. I learnt not to walk by water in a thunderstorm unless you want to get fried.

Any organic molecules that existed would have been fried if they were hit by lightning.

If one Googles all the different theories proposed by evolution scientists, you can find some great examples of their rabid imaginations at work, which in itself proves none of them have any evidence to support any one theory.

If there was true evidence, you wouldn't have all these differing theories.

Another scientist, Alexander Cairns-Smith at the University of Glasgow in Scotland noted Miller did not create nucleotides, which make up the DNA necessary in forming complex molecules

Cairns-Smith figured there had to be another way. He proposed that mineral crystals in clay could have arranged organic molecules into organized patterns and then miraculously organized themselves. We won't waste our time exploring that bizarre one further.

Another one reckons life began at the bottom of the sea.

There is a never-ending supply of theories, some of these guys should write science fiction novels, they would be very good at it.

Despite whatever experiment a scientist may create in the controlled environment of a laboratory, it's all artificial. It doesn't recreate the big bang environment they claim started it all, and they never will.

Lesson: Never believe what you read on face value, especially when it comes to evolution scientists as they are obsessed with proving evolution is correct because if they can't, they will finally have to admit there must be a Creator, and that would destroy their atheist religion. The day has arrived, your evolution theory has been destroyed as there is undeniable scientific proof there had to be a Creator.

Dean Kenyon, Professor Emeritus of Biology at San Francisco State University and a past evolutionist, saw the light and realised evolution can't work after he was asked how the first proteins could have been assembled without genetic instructions. He went on to become one of the pioneers of the intelligent design movement.

Kenyon wrote;

"If science is based on experience, then science tells us that the message encoded in DNA must have originated from an intelligent cause. What kind of intelligent agent was it? On its own, science cannot answer this question; it must leave it to religion and philosophy. But that should not prevent science from acknowledging evidences for an intelligent cause origin wherever they may exist." (*Of Pandas and People: The Central Question of Biological Origins Page 7*)

Renowned English theoretical physicist, cosmologist, author and director of research at the Centre for Theoretical Cosmology at the University of Cambridge Stephen Hawking, who his peers claim was a genius with a super high IQ, wrote an interesting statement that acknowledges the impossibility of all this happening by chance;

"The universe and the laws of physics seem to have been specifically designed for us". Stephen Hawking

Now you see how modern scientific discoveries are proving how impossible evolution is. This has caused some very intelligent and common-sense scientists to realize the evidence proves evolution is an enormous con.

Some appropriate humor to finish the chapter.

One day a group of Darwinian scientists got together and decided that man had come a long way and no longer needed God. So they picked one Darwinian to go and tell Him that they were done with Him.

The Darwinian walked up to God and said, "God, we've decided that we no longer need you. We're to the point that we can clone people and do many miraculous things, so why don't you just go on and get lost."

God listened very patiently and kindly to the man. After the Darwinian was done talking, God said, "Very well, how about this? Let's say we have a man-making contest." To which the Darwinian happily agreed.

God added, "Now, we're going to do this just like I did back in the old days with Adam."

The Darwinian said, "Sure, no problem" and bent down and grabbed himself a handful of dirt.

God looked at him and said, "No, no, no. You go get your own dirt!!!!"

Atheist Religion and Evolutions War against Creationists

Since the evidence for the Big Bang and evolution is so flawed, why do so many people believe it?

With so many compelling common-sense, concise and scientific facts proving evolution is impossible, why is it promoted so strongly by most academics and taught as a fact in schools, colleges and universities?

Evidence disproving the Big Bang and evolution could not be clearer. Recent scientific discoveries on the complexity of the human body, and members of the animal kingdom is conclusive. Evolution is built on unproven theories.

Why do evolution scientists continue to perpetrate unproven assumptions and theories and mislead people? What is the reason, is there a more sinister agenda?

Yes, there definitely is.

I suspect the following statement will be vehemently denied by promoters of evolution, the scientific community supporting it, and education institutions. It will 'ruffle their feathers', an appropriate term considering the humble chicken proves them wrong.

The movement behind evolution is predominantly driven by their atheist religion worldview.

Their theories are not science. They don't meet established criteria for proven science, as you read in chapter one. They spread misleading, false and biased information which is nothing more than 'propaganda'.

Propaganda means:

propaganda (noun) *the spreading of ideas, information, or rumor for the purpose of helping or injuring an institution, a cause, or a person, ideas, facts, or allegations spread deliberately to further one's cause or to damage an opposing cause*

(Merriam -Webster Dictionary)

Richard Dawkins is one of their leading propaganda agents, a staunch atheist who openly ridicules creationists to further his atheistic religion agenda. Educators gladly give him access to schools promoting his atheist religion, where he tells children to question their parents' teachings, which he should be ashamed of. What right does he have to interfere with how parents raise their children? Schools and education institutions should ban Dawkins and his like-minded cronies.

In Dawkins' interviews, when expounding his endless waffling opinions, assumptions and theories on evolution, he can't help but poke fun at creationists and God. He desperately tries justifying his religion of atheism and unsuccessful attempts to prove there is no God. His website is full of atheist anti-God propaganda.

No one, including Dawkins, can prove there is no God, it's impossible. Academics have been trying to do it for centuries and have always failed.

As you have read, the scientific evidence is clear, there had to be a Creator.

Whilst many like Dawkins deny there is a God because they can't physically see God or personally experience Him, this doesn't prove there is no God.

Dawkins and core evolutionists' agenda is clear. They are spreading their atheistic evolution religion because they don't want to be forced to accept there is a Creator that they are accountable to. They want to be their own little gods, and evolution is their invention to usurp the clear evidence of intelligent design.

Even accepting the silly green Martians from inside a black hole theory is more feasible than evolution as it at least shows an intelligent designer.

Evolution is the atheist's god, it's the foundation of their entire belief system. Despite their claims, the fact is evolution is not scientifically proven, according to long standing scientific principles. Belief in evolution is ultimately based on nothing more than faith in endless differing human invented theories. That's the very definition of a religion.

It's ironic how they mock Creationists, claiming they couldn't believe in God unless they could see Him, yet they believe in the

Big Bang and evolution the same way; they can't see it, yet they believe it.

The atheistic religious movement is real. There are even atheist churches. Brian Wheeler of BBC News Magazine wrote an article, 'What happens at an atheist church' after attending one, here are some highlights from his article.

Atheist Logo

Not many sermons include the message that we are all going to die and there is no afterlife. But the Sunday Assembly is no ordinary church service.

Instead of hymns, the non-faithful get to their feet to sing along to Stevie Wonder and Queen songs.

There is a reading from Alice in Wonderland and a power-point presentation from a particle physicist, Dr Harry Cliff, who explains the origins of antimatter theory.

It feels like a stand-up comedy show. Jones and Co-founder Pippa Evans trade banter and whip the crowd up like the veterans of the stand-up circuit that they are.

But there are more serious moments.

The theme of the morning is "wonder" - a reaction, explains Jones, to criticism that atheists lack a sense of it.

So we bow our heads for two minutes of contemplation about the miracle of life and, in his closing sermon, Jones speaks about how the death of his mother influenced his own spiritual journey and determination to get the most out of every second, aware that life is all too brief and nothing comes after it

The "Church of Atheism" practices their doctrine of mainstream science and evolution.

Evolution atheists are at war against creation. Their spin-merchants like Dawkins spread atheist religion by teaching the earth is billions of years old and there couldn't be a God. As science is their god, they naively believe unproven flawed scientist's claims, taking it all in by faith.

It's illogical for them to claim they can't believe in God because they can't see, sense or experience Him according to their personal criteria and unrealistic view of God.

They didn't witness the Big Bang or stages of evolution, nor have they sensed it or experienced it. Yet they believe all the endless unproven theories and unsubstantiated assumptions despite the overwhelming evidence supporting intelligent design and creation.

They irrationally condemn God, blaming Him for world events based on their personal impression of how He should behave, react and show His love. They arrogantly reason how God. according to their personal opinion, should act.

A well-known poem titled, 'God's Tapestry', is a wonderful analogy on how we should view life's events compared to how God observes them.

It describes looking on the underside of a handmade tapestry where all you see is a jumble of different colored threads, no design, no pattern, just a big jumble.

But when it's turned over, you see a beautifully colored orderly pattern that makes perfect sense.

The analogy is that in life we only see the underside of the tapestry of our life and the world.

Events happen that make little sense at times, causing us to question why it happened and sometimes we don't see the answer.

However, the top side is what God sees. He is forming a perfect pattern on which one day we will see a perfect design where everything fits into place.

Evolutionists only want to see the top side of God's tapestry and can't accept they are only permitted to see the bottom where many things in life make little sense because we don't think the same way God does, or have His wisdom, thoughts and knowledge. Because of our finite minds, we can't see the big picture taking shape, it's that simple.

The unsolvable BIG dilemma Dawkins and his like-minded atheists have is they can't prove evolution is true and can't prove there is no God. Their frustration is no doubt the reason for their many false statements and claims. They are trapped between a rock and a hard place. This is the torment of the atheist evolutionist, a problem they will never solve.

I was watching a Dawkins debate on YouTube where his opponent made a statement. He said Darwin realized many things evolution doesn't explain. Darwin was a theist *(One who believes in the existence of a god or gods)* because he couldn't believe that the awesome cosmos and all the beautiful things in the world came about by mere chance or out of necessity. Darwin said 'I have to be ranked as a theist'. Dawkins abruptly interjected, *'That's just not true, it's just plain not true'* (his exact words).

His opponent replied. 'Yes, it's on page 92 of his autobiography. Go ahead and look'.

Dawkins was wrong but this was a classic example of his tactics. When outclassed by an opponent familiar with Darwin's books, he refuses to say a word. This is typical Dawkins. He makes statements and assumptions, knowing his typical audience probably won't challenge him. Evolution spin doctors must be held accountable for their verbal rhetoric that is full of misleading statements, claiming its proven science when in truth they are promoting their sinister atheistic religion of unproven theories.

Evolutionist's arrogance

In the atheist evolutionist's frustration over selling a profoundly flawed theory they rudely and disrespectfully refer to creationists as ignorant or fools:

... we can see that evolution is true. Anybody who is not ignorant, or a fool can see that evolution is true. (Dawkins - PBS interview in faith and reason)

Dawkins, in his classic deceptive spin, uses comments such as *"we can see evolution is true"*. Perhaps they see it in their delusional minds, but that's where it stops. Their spin doctors work on the theory that if people hear untrue statements frequently enough, they will eventually believe the lie. If anyone is smart enough to see through the con, Dawkins belittles them by calling them ignorant or a fool. A stunning claim when in fact he doesn't truly believe it himself and knows it can't be proven.

He belittles anyone who doesn't believe in evolution, and despite his claims that it's true, when cornered in an interview he admitted they don't know. Below is the all-time classic interview showing the staunchest evolutionist admitting they really don't know how it all started.

Richards Dawkins interview with Ben Stein

Stein: *How did it get created? (Referring to the Big Bang and evolution)*

Dawkins: *By a very slow process*

Stein: *Well, how did it start?*

Dawkins: *Nobody knows how it started, we know the kind of event it must have been we know the sort of event that must of happened for the origin of life.*

Stein: *What was that?*

Dawkins: *It was origin of the first self-replicating molecule.*

Stein: *How did that happen?*

Dawkins: *I told you we don't know.*

Stein: *So, you have no idea how it started.*

Dawkins: *No, nor has anybody else.*

Stein: *What is the possibility intelligent design might turn out to be the answers to some issues in genetics or evolution?*

Dawkins: *It could come about in the following way, it could be that in some earlier time somewhere in the universe a civilization evolved by probably some kind of Darwinian means to a very very high level of technology and designed a form of life that they seeded onto, perhaps this planet. Now that is a possibility and an intriguing possibility. And I suppose it's possible that you might find evidence of that if you look at the details of biochemistry molecular biology, you might find a signature of some sort of designer and that designer could well be a higher intelligence from elsewhere in the universe. That higher intelligence would itself have had to of come about by some explicable or ultimately explicable process, it couldn't just have jumped into existence spontaneously and that's the point.*

Do you see the incredible hypocrisy in that last statement? Despite Dawkins' rhetoric that evolution is true and is science, he readily admits nobody knows how it all started, including him. Hello, and these guys claim they are smart. They build an entire argument around evolution being true without knowing the answer to the most basic fundamental question upon which everything else in their theory is built. How it started is the foundation. If you don't know how something started, everything else you claim thereafter is flawed.

It's like me claiming I built a space rocket in my backyard to fly me to the black hole where those green Martians live. You can ask me questions about how it operates and how I will get there, and I can answer those questions as I know theoretically how the rocket should function. However, if you ask to see the actual spaceship as evidence that what I claim is true and I refuse to show it to you, you would rightfully get very suspicious, wondering if what I'm telling you is true.

You would then decide to check out my claims for yourself. When you drive past my backyard, you notice I don't have a spaceship parked there, instead all you see is a pile of metal junk I'm trying to build into a spaceship. So now you know everything I've told you is a lie. You would rightfully never trust me again.

And that's exactly what their theories of evolution are, a pile of junk.

Did you also notice that Dawkins admits there could be a higher intelligence from somewhere in the universe who was responsible for creating life on earth? Its dumbfounding how these guys spend their lives telling the world the Big Bang and evolution is all true and science has proven it; when in truth they don't believe it, knowing the basis of it is flawed and can't be proven. To admit there could in fact be a higher form of intelligence and Creator speaks for itself. That is, provided God doesn't fit into that scenario, as that destroys the atheist's religion. How hypocritical.

Did you also notice he mentioned that evolution all started with the first, *self-replicating molecule.* How did this first molecule form so it could self-replicate, no answer to that one.

In a Feb 28, 2014 interview Dawkins made this statement:

"Genes are immortal in the sense that the coded information they contain is reproduced replicated with almost total fidelity significantly not absolutely totally fidelity generation after generation after generation such that there are genes which are identical to what they were tens of millions of years ago, 100 of millions of years ago in some cases....so genes are immortal"

You read about genes in chapter six, so is Dawkins' statement true. No.

To refresh your memory, this is what you read.

Genealogists use DNA to verify two individuals are closely related as a person's DNA gets passed on to the next generation. DNA is passed down to the next generation in big chunks called chromosomes. Every generation, each parent passes half their chromosomes to their child. Most people have 23 pairs of chromosomes for a total of 46. One of each pair comes from mom and the other from dad. This is why we are 50% related to our moms and 50% to our dads.

Genes contain DNA, a child gets 50% of both parent's chromosomes. When that child has a child, your grandchild, it gets diluted further to where only 25% of their genes are related to you, and so on.

If nothing happened to the chromosomes between generations, there would be around a 1 in 8 chance that you would get no DNA from a great, great, great grandparent.

Dawkins' claim that genes are immortal is not true, genes change with each generation.

To illustrate the craziness of his claim, how could he ever prove the genes of today were identical to genes tens of million years ago, or hundreds of millions of years ago. Where is the evidence to prove it? Where are the samples from 100 million years ago to authenticate his claims? Obviously another unprovable claim. They can't be proven to exist unless a scientist 100 million years ago had the foresight to store some of them in a safe.

As Dawkins claims he evolved from a fish, perhaps he is an exception and is a 100-million-year-old evolutionary mutation reincarnated with the same genes, which explains why so much of what he says sounds very fishy!

Can you see how ridiculous their claims are? And to think they continually pump out this rubbish, making ridiculous, unproven and impossible statements which can't be proven. To use Dawkins own words… 'That's just not true, it's just plain not true'.

Pope Benedict XV1 even commented: 'Richard Dawkins, *The Selfish Gene* book is a classic example of science fiction'. Exactly what Dawkins writes, it's just science fiction.

The negativism of Dawkins and evolutionists.

It, (Darwinism) provides the only satisfying explanation for why we exist, why we are the way we are. - Richard Dawkins.

How sad and depressing. Evolutionists teach we merely exist as part of the evolutionary ongoing random mutation process, and who we are and why we are here is just the result of mutations in DNA.

If this is the world according to Dawkins and evolutionists, there is effectively no hope for humanity and possibly a contributing factor to the attitude of many young people who have been taught evolution in our education system. Sadly, they are being taught false and misleading information.

What a depressing thought for young people, DNA is the only reason you are who you are, and it affects who and what you will turn into. This is how harmful evolution is and why it's so vital we drive this negative atheist religion out of our educational institutions. Students need meaningful constructive character-building principles to give them hope.

Teaching students they are simply a random mutation of evolution and their destiny has been determined in their DNA by the mystical force of evolution is uninformed, ignorant, disrespectful, and harmful to youth.

Students should be taught positive creationism; that they have a free will and can be anything they want to be.

Study, work diligently, and you can achieve your goals in life, thus leading to a rewarding and fulfilling life.

In Dawkins' Watchmaker book, he takes yet another arrogant shot at USA Christians; *"there are disturbing signs that their influence (Christians) is ever growing, at least at local areas of the United States. Insofar as the backwoodsmen have arguments, they are mostly centered around the notion of design."*

According to Dawkins, if you believe in intelligent design and a Creator you are a *"backwoodsman"*, how's that for atheist academic arrogance and disrespect.

In his book Dawkins continually makes unsubstantiated claims; *"our existence once presented the greatest of all mysteries, but that it is a mystery no longer because it is solved. Darwin and Wallace solved it"*.

Absolute rubbish is the only fitting term for that garbage. It is just another of the many untrue Dawkins statements. Darwin and Wallace proved nothing; it is just more of his spin in a desperate attempt to further his atheism religion.

The following statement from the same book demonstrates how delusional and misleading Dawkins is.

"More, I want to convince the reader, not just that the Darwinian world view happens to be true, but that it is the only known theory that could, in principle, solve the mystery of our existence."

He is so delusional, believing a 'theory' to be truth, and simply a theory is all that's required as proof of the mystery of existence. The green Martians coming out of a black hole and creating earth is also a theory.

Anyone can create a theory, but that's all it is until proven, and despite what Dawkins claims, evolution has NOT been proven, it never will be, it's impossible to prove.

Here is another classic untrue Dawkin-ism to make you feel like a better person; your ancestors were fish.

"You can go back as much as you want. I propose going 185 million generations ago; you will realize that your ancestor of 185 million generations was a fish" (Richard Dawkins)

How can anyone go back 185 million generations! There is absolutely NO EVIDENCE or proof showing a transmutation of fish into humans. It's outrageous, unbelievable rubbish he continually spews forth.

Next, he might suggest we avoid eating fish as we might unknowingly eat a relative, an uncle, or cousin.

This nonsense is endless, I could fill hundreds of pages but I'm sure you get the picture.

Evolutionists also deny we have a spirit.

Do you believe you have a spirit?

This is an extremely important question and a major flaw in evolution.

Evolutionists can't explain where the human spirit and our ethics or moral code came from. By moral code or ethics, I don't mean laws of the country. Instead, I'm referring to a person's inner belief on what is right or wrong, as evidenced by their conscience.

We read above that Dawkins and evolutionists claim we originally came from fish who evolved into animals. So, if we come from animals, why don't we survive based solely on animal instincts? Where did humanity's moral code come from?

Philosophers who study human behavior in search of answers on why people do certain things and the meaning of life come up with many theoretical ideas, according to their outlook on life. These philosophers have been around for a long time, churning out an endless number of books and papers expressing these opinions. The meaning of life is one of their favorite subjects, one that has troubled them since they invented philosophy.

History is littered with 'Philosophers' and their writings. Socrates, who lived around 300 BC was possibly the most famous of this group. He sought to draw his followers into thinking about questions of life through a series of questions. They continually question and propose vague and abstract ideas. Interestingly, he was found guilty of 'impiety' and corrupting the young and condemned to death by drinking poison.

Evolutionary scientists love philosophers, as they join in on evolutionist's abstract ideas. It's a contest to see who can invent the craziest idea. Any theory from one's imagination they can verbalize convincingly so a few of their peers also agree is the only criteria.

Where do human's morals and principles come from, who sets the standards for truth?
Morals and principles are not found in genes or DNA. Fish don't have them, so it couldn't be passed down from our fish ancestors, as Dawkins claims. How we view things in society isn't determined by our DNA.

Humans have a free-will, we can decide what action to take in an event or situation. That's obvious, as sometimes we make good decisions while at other times we make not so good decisions, resulting in unpleasant consequences. Everyone can relate to that. For every action, there is a reaction, that's a law of life.

Evolutionists claim we are nothing more than a by-product of evolution, and our genes ultimately determine who we are and our destiny in life. Somehow this magical, invisible, uncontrollable force of evolution has predetermined what type of human being we become. How can that be when there are no morals or principles in genes or DNA code. Evolutionists cannot deny this, as they can't produce any evidence of DNA moral coding.

Evolutionists claim that how a human develops, be it a good person who is kind and generous and an asset to society; or a bad person who may steal, cheat, lie and kill someone, is predetermined in their genes, which are governed by the law of evolution.

If the force of evolution dictates an individual's moral code as evolutionists claim, where do those moral codes come from? How have they been built and on what basis? Who or what was the original authority of the moral codes? Was it from fish, dinosaurs or animals?

What are the morals and principles of evolved atheist evolutionists? What moral code do they adhere to?

These codes are not to be confused with a government's laws based on longstanding Biblical commandments. We know there is a punishment if you break them, which deters many people, knowing they might end up in jail.

We know staunch evolutionists must be atheists, as the basis of evolution requires a denial that God exists, and that He created the universe. E.g. Dawkins and his like-minded evolutionists openly deny God exists and are atheists. I have listened to many of his and other leading evolutionists' lectures and debates in universities promoting evolution.

He mocks the existence of God while his audience of fellow atheist evolutionists laugh at his derogatory opinions of God.

If they openly deny God exists, then they must also reject any moral code or standards set by God.

I appreciate this may not apply to all who believe in evolution, as some sit on the fence with one foot in each camp, believing in a God whilst also believing in evolution. I have to confess, I'm not sure how they rationalize that, as the basis of evolution denies a creator and the Biblical account of creation. Hopefully, this book enlightens them to the truth.

Excluding the laws of the land set by governments, what moral standards do evolutionists follow since they have no sovereign intellectual supreme authority.

If I was an atheist evolutionist, why can't I lie, steal, cheat people, kill, or do whatever I want, as my DNA evolved from animals and that's where my moral standards evolved from.

Darwinism teaches it's all about survival of the fittest, and this is the atheist evolutionist's bible.

According to evolutionists I evolved from animals, so I still have remnants of those genes effecting my decision making. Excluding the laws of the country telling me it's wrong to kill, who says it is wrong? That's only the views of lawmakers, who says they are right, it's just their opinion.

If my evolved animal DNA tells me it's ok to kill for survival, who are you to tell me I can't? As an evolutionist, my claim is supported by science and it is only natural to follow my ancestral animal instincts.

If this is sounding a bit weird, don't worry. This is one of those winding roads with a blind corner you're about to go around, so fasten your seatbelt as this is building to a major point.

As an atheist evolutionist I have no moral conscience established by a higher authority telling me what is right or wrong. If I can do the crime without getting caught by police, I see no problem with this as my moral code established in my genes through evolution predetermines how I think and act.

If another person thinks it's wrong, who are they to judge and tell me it's wrong? It's just their opinion based on their genes and how evolution predetermined their views on what might be right or wrong for them.

Evolutionists claim the mystical force of evolution has predetermined that I just happen to be living in this moment in time, and the evolution of my genes sets my course in history.

Do you see the very dangerous underlying moral dilemma created by the theory of evolution? Evolution has no authoritative base so it has no absolute moral code.

In the atheist evolutionist's world, there is no moral law or authority, as compared to creationists who believe in God and endeavor to follow His laws and the moral standards He established for humankind's own good. To the creationist, God is wisdom and truth. He established their moral laws and standards; He is their ultimate authority.

To the atheistic evolutionist, there is no ultimate authority that sets moral standards, there is no source of truth. Whatever they imagine truth is... becomes truth to them. They can set their own moral and ethical standards.

They can say evolution is true, it's proven; and as they have no moral benchmark set by a higher authority, there is no standard of truth or perspective of truth in their world. After all, to them what are the ultimate consequences if they go through life lying and cheating? There are none, whereas creationists know they are accountable to an ultimate authority who sets moral standards to live by for humanity's good. We know that doesn't always happen, as we have a freewill.

Moral and ethical standards are another catastrophic flaw in evolution and the atheist's religion which is flowing through society. Moral standards and the definition of truth are rapidly being eroded, even at the highest levels of government.

Dawkins claim that our genes drive us and determine who we are is a load of rubbish. We all know we have freewill; we can decide to rob a bank or steal a car. Freewill enables us to make decisions based on our moral code, and that's the key; it's based on one's moral code. A strong moral code tells us not to do it, a weak moral code says who cares, if it feels good, do it.

If a person follows their natural built-in moral code established by the Creator, they hopefully won't make the choice to rob or steal as their conscience tells them it's wrong.

That's how the Creator wired us, that's why we have a conscience.

If a person continually suppresses listening to their conscience, it will eventually diminish, and they may not have any compunction about lying, cheating or stealing, as their moral code allows it.

For Dawkins to claim that what we decide is in our genes, is intellectually irrational and irresponsible. Dawkins and his like-mined cronies clearly haven't thought through the long-term implications of their atheist religion of evolution and its long-term morally eroding effect on society. A criminal could claim in court it wasn't his fault, it was just his genes directing him to steal or kill, so he had no choice. After all, science has proven it.

Evolution theory is a major contributing factor in the breakdown in respect for authority, laws and moral standards as their teaching has no central source of authority.

To evolution atheists, truth is whatever an individual chooses it should be, truth to them is what they wish it to be.

Dawkins is a perfect example of an extreme atheist evolutionist, so many of his statements he claims are true, proven and factual are not, they are just unproven theory.

In his own 185 million generations fish-evolved brain, his claim not mine, Dawkins may well think his claims are truth based on his moral standard of truth. Perhaps atheist evolutionists genes have evolved to a stage where they assume anything dreamed up in the name of science is truth.

Perhaps they are just victims of the mystical, uncontrollable force of evolution that has shaped their genes and they are not responsible for what they say, as they are under 'the force of evolution'. The force is with them, controlling what they say and do, therefore they are not accountable to anyone. It's all the power of 'the force' at work. The delusion of evolutionists.

Atheist evolutionists are living in a delusional world based on unproven theories.

How do atheist evolutionists explain the spiritual realm?

Atheists don't believe in a spiritual realm; they believe it's a trick of the mind, it's nothing more than the brain playing tricks.

If you believe you have a spirit and there is a spiritual realm, you cannot believe in evolution and atheism as they deny it exists. They have no explanation for it, other than arrogantly claiming it doesn't exist. Evolution scientists can't put the spiritual realm into a test tube, so they naively claim it doesn't exist.

In a 2017 Gallup poll, 87% of Americans said they believe in God or a higher power. That's faith in a spiritual realm, that's believing in a being they can't physically see.

Christians quote an interesting Bible passage to infer the Creator has implanted into every human being a spiritual awareness of Him at birth, it's part of a person's spirit.

> *For even though they knew God [as the Creator], they did not honor Him as God or give thanks [for His wondrous creation]. On the contrary, they became worthless in their thinking [godless, with pointless reasonings, and silly speculations], and their foolish heart was darkened. Romans 1:21. (Amplified Bible)*

Based on this verse, it's up to every individual to build on that awareness and either develop a relationship with God or reject Him, which atheists do. Their rejection doesn't prove there is no God, it just means they have personally decided they don't want to be accountable to a higher authority, that's their choice.

The Creator gave everybody a free will to decide if they want to learn more about Him, or deny He exists; blocking out any awareness of Him or the spiritual realm. This is what atheists do.

They made a conscious decision to block out God as they can't accept there is a higher authority possessing a far superior intelligence than them.

We consist of a body, soul and spirit. Every common-sense person knows they have a spirit, it's who they are. This is even evident in some of our frequently used sayings. We say, 'that child is spirited', or 'we have a spiritual connection'.

The existence of a person's spirit doesn't need to be scientifically proven; every human knows they have a spirit within themselves, it's the inner person. Atheists simply block this out by claiming it doesn't exist because they don't have an answer for how it evolved.

Let's use the common-sense test to determine if we have a spirit. Here are two simple examples.

Why is it that we sometimes meet people we feel comfortable with, there is just something that clicks. Then there are other individuals we meet and we just don't seem to be able to get on with. There is a clash, we just don't feel comfortable with them. I'm not suggesting there is something wrong with those people, they may be very nice. It's simply there is no common- bond, that is a spirit thing, our inner spirt doesn't click with that person.

Why were you attracted to your husband/wife, partner, girl/boyfriend? Initially, you may have said they looked great. There is a physical attraction, however for any relationship to last there must be a closer bond. There is something inside us that bonds with that person, it's a spirit thing. I recall when I first met my wife. there was something about her that attracted me.

Sure, she was physically attractive but there was also her personality and spirit. There was something much deeper, our spirits connected.

We have been married for many years and survived numerous storms of life that have drawn us even closer. Sometimes we can glance at each other and know precisely what the other is thinking, that's a spirit thing. It has nothing to do with our brains, as we haven't expressed our thought at that point. Then, after one of us says what they were thinking, the other comments how they were thinking exactly the same thing at precisely the same moment. That's a connection of our spirits, that's a spiritual realm, we all have a spirit, it's common-sense.

If we were just random mutations without a spirit, there would be no relationships. We would be no different than the animal kingdom. Even saying that, I believe they also have a spirit. I had a beautiful dog that definitely had a spirit. She was loyal, trusting and had feelings. When caught ripping up the garden, I just looked at her and she ran off to hide. She sensed my spirit and knew she was in big trouble; animals definitely have a spirit.

The teachings of atheist evolutionists that humans have evolved from animals with no soul or spirit is morally corrupt.

It teaches humans have inherited their primeval ancestor's animal survival instincts, kill or be killed; do whatever you have to do to survive for that is all that matters. Remember, their high priest Dawkins claims genes don't change, they are passed down for hundreds of millions of years.

Do you see where the evolutionist's philosophy is taking society by fueling our moral decline.

For evolutionists to claim there is no spirit component to a human is ridiculous. It's one of the dumbest things I have ever heard, any common-sense person knows that is not true.

There is a publication, *The Atheist's bible*, a VERY depressing read. Don't read it unless you want to ruin your day. It has no hope, no future. Anyone who reads it would have to question the very purpose of their existence.

That's NOT what life is about, and if anyone happens to read that rubbish and finds themselves in that space, let me advise you get out of it as quickly as you can. Atheism religion is bad stuff, it has no hope or future.

Here are some examples of their depressing doctrine.

... we are mainly the product of lots of random mutations. So, the fact that I am writing this book, and that you are reading it, is largely due to a combination of an incredibly large number of random factors. So, why is it us who made it into the 21st century? Why did life evolve here, and not somewhere else? Why was it our species that developed reading, writing, and thinking? The answer to all these questions is the Anthropical Principle. It says: If someone asks such questions, then all the conditions for life and intelligence have been met in this place at this time for this species.

If you have never heard of the 'Anthropical Principle', apparently it means:

'the cosmological principle that theories of the universe are constrained by the necessity to allow human existence'.

They claim there is no reason we exist other than our parents' desire to have sex. We exist because we were born; we live, and then we die and go back into the earth. That's it, according to atheists.

That's all you have to look forward to, and after you die, there is nothing. According to them, that's your lot in life.

I struggle to grasp why anybody would want to believe in an atheist's evolution religion. It has no hope, it's illogical. Why would you believe in theories teaching you are only a random mutation? You live out your life as one tiny speck in the process of an infinite number of random factors. Then you die, and that's the end. They put you in the ground and you go back into the earth and become worm food. It's the religion of no hope, how depressing.

On this fact alone I would believe in a Creator God any day, it's the most logical explanation after we understand the amazing complexity and structure of the animal kingdom and the human body. Just take DNA on its own, there has to be an intelligent designer, all other options are illogical. The chicken was created, no other explanation.

Creator God gave us a body, soul, and spirit, we were created for a purpose. We have a freewill and can be the person we want to be. When we die our spirit and soul lives on and goes to a spiritual realm. That's hope, that's something to look forward to, that's common-sense to believe in a faith of hope and future.

I want a piece of that action.

Below is the atheist's depressing doctrine on 'the purpose of life'.

Now if we want to find the purpose of our life, we have to find someone who pursues some intention with us. Then we ask that person for that intention, and this is our purpose of life. We have already seen that, in the atheist world view, God cannot take this role. Now who else could pursue an intention with us?

How about yourself? Do you have an intention for your life? If yes, then this is the purpose that you have given to your life. If your intention is to make money, then making money is the purpose that you are giving to your life. If your intention is to be happy, then being happy is the purpose that you have given to your life. Congratulations, you now have a purpose of life! It is, quite plainly, whatever you choose it to be. You are the one who chooses what to do with your life.

This is not a particularly smart insight. It just follows from the definition of the word "purpose": your body is but a tool, and the purpose of a tool is whatever intention someone pursues with it — on this occasion you yourself.

But maybe you do not pursue an intention with yourself. Then don't worry! If you refuse to give your life a purpose, and if you keep seeking for a purpose, then you will eventually find someone who will offer you a purpose for your life! Plenty of people are happy to pursue their intentions with you. They will tell you what is your purpose of life! In other words: If you don't know what to do with your life, then others will tell you what to do with it.

The problem is that what others tell you is often not in your interest. So, you better give your life a purpose by yourself.

As with most atheist doctrines, they are confusing as they go against common-sense logic. They make little sense as it's not how humans were designed to function.

They claim the purpose of life is to find someone to pursue some 'intention' with, I'm guessing they mean a plan or aim. A common-sense person would say 'a life partner'. Then you come up with an intention for your life like making money; that's deep!!!

They must have forgotten making money is part of everyday life.

If we want to eat, have a car, house, travel, look after our children, we all need to make money.

The other alternative is if you don't need money just have an intention to be 'happy'. That may be hard if you don't have any money to live off.

The best they can offer is, 'you just need to give yourself a purpose.'

If you're an evolution atheist, please answer this question.

How can you have your own purpose? Your god, 'evolution', that mysterious random force with no brain or plan, yet is supposedly responsible for everything around us through billions of random mutations, has destined who you are and what you will do?

See how confused and mixed up atheist evolutionist teaching is. They invent whatever suits them. Without belief in a soul and spirit all they have is emptiness, a depressing doctrine with no hope.

When reading atheist evolutionist's literature, you will see a pattern showing why they believe evolution. It's the simple rejection of the authority God represents. They want the freedom to do what they want and don't want to be accountable to a higher authority.

They like evolution and atheism as it doesn't teach any morals; you can do whatever you want, you can be your own god, thus providing a more convenient belief system compared to believing in the Creator.

Some also can't accept the principle of a higher supremely intelligent authority.

They want to reason God by putting Him in a test tube to analyze. Where did He come from? His IQ, or whatever test they consider in their world would prove God exists. They claim they can only believe in God if He appears to them on their terms.

They fail to understand it doesn't work that way. God is under absolutely no obligation to meet their demands and prove himself to him based on their criteria. Because they believe in evolution, which denies any spiritual realm, they have willfully tuned out his voice. Like a radio, you tune into the station to hear and open your ears to listen. The ears of their spiritual dimension are closed.

God works in a different realm, which they don't grasp or don't want to accept. That's their choice in life, and that's why God gave everyone a freewill to make that choice.

Many atheist evolutionists use the old 'tyrannical deity' argument as their justification for not believing in the Creator. They claim that previous religious wars were God's fault, or if He was God, He could stop them.

They don't understand that humans are the ones who create religious systems in the same way they have created their religion of evolution atheism. Humans have been responsible for many atrocities over the ages and despite being called religious wars, they were nothing more than humans conquering other humans.

While citing religious wars like the crusades, they conveniently ignore the estimated 60-80 million killed under Hitler, Stalin, Mao, and the Khmer Rouge, all staunch atheists who wholeheartedly followed and put into practice Darwin's principles of survival of the fittest.

Radical evolution atheists are at war with Creationists, wanting to convert them and their children into atheists.

God didn't create any corrupt religious system, mankind creates religious systems, evolution and atheism are just another religion.

The Creator gave individuals a freewill to believe and follow what they choose. We make our own choices on who and what we believe in, which ultimately decides one's destiny.

Two guys who didn't know each other were relaxing on a park bench at the highest point of a small hill overlooking a large, expansive and beautiful park. One was older with grey hair, while the other was much younger. It was spring, lush velvet green grass rolled out across the landscape, encircling a serene blue lake. The area was home to an array of wildlife, ducks, geese, herons and swans. Children were feeding the birds and others were contentedly playing by the lake edge, laughing, running, and forgetting life's challenges as they escaped into the serenity of nature.

The older man, Bernie, who believed in God spoke up, "Isn't Gods creation wonderful".

The younger man, Chris, responded, "I don't believe in God, I'm an atheist".

Bernie replied, "Why would you want to be one of those".

Chris replied, "Show me God and I'll believe in Him, I can't accept someone I can't see. Besides, he tells me how I should live, and I don't want anyone telling me how to live".

Bernie contemplated Chris's answer for a few minutes before saying, "I look at it this way, I believe in God. Look at this magnificent scenery and the miracles of nature. There had to be a designer and Creator, and I believe evolution is impossible. It doesn't make sense, and even better, when I die I know I will go to heaven for eternity".

Chris scoffingly laughed, "Prove to me there is a heaven".

Bernie replied, "And you can't prove to me there isn't a heaven or a hell. I have studied evolution and discovered it takes more faith to believe in an unproven theory created by atheists than to believe in a Creator. I also have the added guarantee that, if I'm wrong I have nothing to lose because I'll end up where you end up. However, if you're wrong and there is a Creator God who you rejected because you wish to live your life how you want, then you have so much to lose for eternity. It's a no brainer to me to follow a faith of hope and a future".

Chris replied, "I see your point, I'll certainly have to give that a lot more thought".

Epilogue

Your journey of discovery has almost reached its destination. Along the voyage thus far your eyes have discovered many compelling facts disproving the theory of evolution, how atheist evolutionists present theories as if they were scientifically true, and you have read examples of their deceptive and misleading rhetoric from their own words and writings.

On this adventure you have discovered how modern science not only proves evolution impossible but shows there had to be an intelligent Creator, these facts are proven using demonstrable evidence.

I have presented examples of leading evolutionists promoting the atheist agenda by aggressively pushing their false religion of evolution.

The stream of unproven, misleading and wacko ideas continues to increase as they realize they can put forward any theory, inferring its true, knowing they will never be called on to prove it.

By hiding behind the real scientists who came before them like Newton, Pascal, Volta, Tesla, Hooke, Galileo, Bohr, Marconi, and others, they have convinced the public to innocently believe their theories are plausible and reliable when they are not; for in the evolutionist's world, truth is subjective. There is no moral code in evolution or the atheist religion.

Unfortunately, evolution scientists and educators, especially those in the universities, have an agenda. By teaching the false theories of evolution, it naturally results in no longer believing in a Creator, which was a core foundation of science for centuries. The foundations of evolution deny the existence of a Creator, therefore if they truly believe what they are teaching, they must logically be atheists.

For anyone to claim a belief in the Creator and evolution at the same time is an irreconcilable contradiction; you can't believe in both as evolution at its core denies a Creator.

Evolution scientists actively promote atheism, using evolution as the basis of their belief system and their bible; it is the key sacrament in their religion.

Evolution has no benefits for mankind. It has no moral authority or substance, with no basis of truth and social standards as they deny a sovereign intelligent Creator who has established society's moral and ethical principles.

As atheist evolution teaches no moral basis, the expansion of its teaching has been the major contributor to the rapid erosion of the moral code society has held for centuries.

Over the past 50 years, there has been a direct correlation between evolution being widely taught in the education system and the decline in moral standards, lawlessness and disrespect for authority we see so prevalent today. Up until the 1960s and even early 1970s, in most places in America you could leave your house unlocked with no fear of being robbed.

Teaching the theory humans evolved from fish ancestors, and your life is just an insignificant part of a random, unplanned process, providing no future hope beyond this life is wrong and unproven. Atheist evolution is depressing, morally destructive, and counter-intuitive to the social fabric of society and humanity.

The evidence now shows that the teaching of morally devoid evolution theories by educators aggressively promoting their religion of evolution in schools and universities at the exclusion of biblical values and creation has been a major contributing factor in the breakdown of moral standards, resulting in rampant drug abuse, youth violence, school shootings, a general lack of respect, and the very sad and alarming increase in youth suicides.

Today's public education system is devoid of teaching the core moral standards which come from the Creator for the purpose of society functioning in a caring, productive way.

The numbers of youth suicides in USA alone is alarming.

*Between 2000 and 2007, the suicide rate among youth ages 10 to 24 hovered around 6.8 deaths per 100,000 people. Then, the rate curved upward, reaching a rate of 10.6 deaths per 100,000 by 2017 — **a 56-percent increase in less than two decades**.*

(PBS, Oct 18, 2019 - Youth suicide rates are on the rise in the U.S.)

With more than 6,200 suicides among people aged 15 to 24, suicide ranked as the second-leading cause of death for people in that age group in 2017, trailing behind deaths from unintentional motor vehicle accidents, which claimed 6,697 lives.

(PBS, Jun 18, 2019 - Suicide among teens and young adults reaches highest level since 2000)

These figures bear out a direct correlation to the ever-increasing aggressive stance from educators teaching evolution while aggressively challenging students who believe in Creation.

Here are more sad statistics.

Suicide is the SECOND leading cause of death for ages 10-24. (2017 CDC WISQARS)

Suicide is the SECOND leading cause of death for college-age youth and ages 12-18. (2017 CDC WISQARS)

More teenagers and young adults die from suicide than from cancer, heart disease, AIDS, birth defects, stroke, pneumonia, influenza, and chronic lung disease, **COMBINED.**

Each day in USA, there are an average of over 3,069 attempts *by young people grades 9-12. If these percentages are additionally applied to grades 7 & 8, the numbers would be higher.*

There is a "Silent Epidemic" sweeping through our nation that claims an average of more than 100 young lives each week

(The Jason Foundation, Inc – JF The Parent Resource Program)

According to the Centre for Disease Control and Prevention's 2017 Youth Risk Behavioral Survey, over ONE out of every FOURTEEN young people in the USA attempted suicide during the previous 12 months.

My wife and I were at lunch with a friend in her 70s. During our casual conversation, I happened to mention I was including a correlation between the rise in atheist evolution taught in our education system and suicide in this book, quoting the statistics. She commented that during her school days she never heard of students thinking of committing suicide. My wife and I immediately pondered and reflected back to our school days and realized we agreed with her. Neither of us could recall any youth suicides or depression, and that wasn't all that long ago.

Evolution was not taught or spoken of in the public schools when we grew up. Instead, Christian education was taught and the youth had respect for authority and for themselves. Drugs didn't exist, the crime rates were low, and there was certainly no school shootings. In America, even in the 1970s, it would not be uncommon to drive through a high school parking lot and see rifles in the racks of the pickups of many teenagers. Youth had respect for other's views even if they disagreed with them, unlike what we see today on liberal campuses, which are the fertile breeding ground for the breakdown in social standards.

Youth suicides and depression on the scale we have today only appeared in the last 20-30 years. Not coincidently, this was the same timeframe when evolution began being progressively introduced into the education system.

This alarming suicide rate isn't just confined to youth. A January 24, 2020 CDC report disclosed the US suicide rate of persons aged 6-64 has increased by 40% in under two decades. In 2017, nearly 38,000 persons committed suicide.

Evolution scientists and educators always vehemently deny any link between teaching on evolution and suicides, inventing another of their many theories to counter these claims; however, you be the judge. The evidence speaks for itself.

Youth are now taught evolution and atheism, the depressing religion and belief system that promises no future beyond this life. No moral codes, ethical principles or respect for authority. It's a no win, no hope, morally devoid and corrupt religion built on unproven theories. A loser religion.

History shows that previous cultures self-destruct when a nation no longer has a sense of purpose beyond just existing. Evolution gives no purpose, so it is no wonder that an alarming number of youths are self-destructing. Lawlessness is rampant, again this is the logical progression when Darwinism is put into practice in a society.

What benefit is there in teaching students an unproven socially destructive theory devoid of purpose, other than fulfilling the agenda of atheist evolutionists and educational institutions who vehemently want to remove the Creator from society? These individuals bear a very large portion of the responsibility for the breakdown of moral standards in society.

Evolution gives youth no hope, it's a depressing, soul-destroying religion with no future. They should ban it from schools and universities.

If any young people are reading this, please hear me when I tell you evolution is not true. You are far more precious than some supposed random mutation. Instead, you were created by an incredible Creator who brought you into this world for a very important and specific purpose that no other person can fulfil. Every human being is very special, you are a wonderfully and uniquely created individual with a purpose in life, and your life doesn't cease when you get put in the ground. The advocates and teachers of evolution have lied to you, and they should be ashamed at what they are doing not just to society, but to precious individuals like yourself.

Mainstream Universities are now a cesspool, completely overrun with the academic atheist evolution science educator's intent on driving any belief in a Creator from the minds of the students placed under their care.

The following is just one of many examples:

"There is an insidious and growing problem," said Professor Jones, of University College London. "It's a step back from rationality. They (the creationists) don't have a problem with science, they have a problem with argument. And irrationality is a very infectious disease as we see from the United States."

Professor David Read, vice-president and biological sciences secretary of the Royal Society, said that they felt it was essential to address the issue now: "We have asked Steve Jones to deliver his lecture on creationism and evolution because there continues to be controversy over how evolution and other aspects of science are taught in some UK schools, colleges and universities. Our education system should provide access to the knowledge and **understanding gained through the scientific method of experiment and observation, such as the theory of evolution through natural selection**, *and should withstand attempts to withhold or misrepresent this knowledge in order to promote particular beliefs, religious or otherwise." (The Guardian, Feb 21, 2006)*

Here we see a professor teaching a theory as if it was truth. This man needs to get his dictionary out and look up the meaning of 'theory'. He also needs to open his eyes to biological science, which continually provides a mountain of evidence proving evolution is impossible. That is true science based on the long-standing definitions, but they ignore clear scientific evidence when it disproves evolution.

It is fascinating how the word of God, predicted this would happen nearly 2,000 years ago when the Apostle Paul warned, " *O Timothy,* **keep that which is committed to thy trust**, *avoiding profane and vain babblings, and* **oppositions of science falsely so called**:" (1 Timothy 6:20 KJV).

Note how the evolutionists and atheists label anyone who doesn't accept evolution as "irrational". They are hypocrites, as they teach irrational unproven evolution theories, claiming they are acceptable to teach simply because science promotes them.

The hypocrisy is astounding, it's ok for them to teach UNPROVEN theories, random observations, and all the other rubbish of natural selection, which is untrue and unproven because it's all done in the name of science. If this is their true benchmark does it apply to all other subjects taught, no.

In a CBN News article, a Yale Professor who turned his back on Darwinism made the following statement:

Gelernter explains, "As far as they (evolution educators) are concerned, take your life in your hands to challenge it intellectually. They will destroy you if you challenge it."

"What I've seen, in their behavior intellectually and at colleges across the West, is nothing approaching free speech on this topic," he continued. "It's a bitter rejection, not just — a sort of bitter, fundamental, angry, outraged, violent rejection, which comes nowhere near scientific of intellectual

discussion. I've seen that happen again and again. 'I'm a Darwinist, don't you say a word against it, or, I don't wanna hear it, period.'"

"I am attacking their religion," he added. "It is a big issue for them."

(CBN, Aug 23, 2019)

This is a real-life example of the evolutionist's war against creationists, it is real and the decline in society is collateral damage from this scorched earth policy of theirs.

Parents and students need to push back against the teaching of this soul-destroying false atheist religion. It's time for government to legislate removal of the theory of evolution in educational intuitions, as it is the religion of atheists and religion is supposedly banned in schools today.

The odds factor

In Antony Latham's book *'The Naked Emperor: Darwinism Exposed'*, he shows the impossibility of evolution.

Latham calculates the odds of the probability of the infinite amount of mutations required in the perfect sequence occurring for the human eye to evolve. Remember, all the mutations have to happen in perfect sequence, like building a car. You can't install the headlights until everything required to hold them in place and provide the energy is also in place and functioning. Not to mention how the light units themselves need to be made perfect to start with.

If there were 10,000 steps needed in the process for an eye to evolve, which is drastically understated, there would in fact be many more; each step in the mutational process of the eye formation must arrive by sheer chance, according to evolutionists.

Assuming each mutation had a probability of occurring in one in 10 million individuals (a very conservative number), then the chance of getting all the 10,000 mutations in series is 10 million times itself 10,000 times. Therefore, the odds of a series of mutations occurring is one chance in 10 to the power of 70,000.

As a comparison, this figure greatly exceeds the number of atoms in the known universe. So, it's totally impossible. It simply can't occur, it's that simple, so why don't evolutionists get it if they are truly as objective as they claim.

Do you hear that evolutionists...it's impossible. This means it's totally irrational to believe in evolution. What don't you understand about those clear facts?

Deceptive evolution scientists tell us they are so smart but refuse to see the obvious, even when it's staring them in the face, why? The simple reason is because they don't want to. It has nothing to do with the evidence, they are willfully turning a blind eye. They are like a person who says they can't see a light while their eyes are closed. You politely suggest they open their eyes, so they open them while covering their eyes with their hands. Then they say, "I still can't see it". Rolling your eyes, you say, "put your hands down and then you can see it." They willingly comply but turn around before lowering their hands. "I still can't see it."

Their agenda is all about proving there is no Creator and God, and they are so consumed with this goal they refuse to see the obvious.

It's so clear, even the humble chicken proves it scientifically, yet they still can't bring themselves to accept the possibility of a higher supremely intelligent Creator.

A person can become blinded by their own thoughts, this is the "disease" which has afflicted the evolution scientists and academics teaching evolution. Their imaginations and thoughts have blocked their grasp of common-sense reasoning.

Academic and aristocratic people live in such an uncommon atmosphere

that common sense can rarely reach them.

Samuel Butler

A closing message

Students: Evolution is a dead end, it's a road to nowhere. There is absolutely nothing to be gained by following its tenets, there are no benefits to you whatsoever. It's a big lie and con, perpetrated by atheist's intent on promoting their atheist depressing loser religion. Biological scientific evidence proves there had to be a Creator.

If you are pursuing a career in science, follow a path worthwhile to humanity such as medicine or engineering.

Parents: My sincere desire is for every student to read this book so they will have available to them an opportunity to critically search, question and understand what has falsely been taught to them in schools and universities. Then, armed with this knowledge, that they will hopefully have the desire to explore more common-sense factual and positive truths about our magnificently created world, knowing they are a uniquely created individual and learn to be proud of who they are and what they can accomplish in life.

Evolutionists: If you have believed in both God and evolution, they are not compatible. Evolution denies a Creator and is the basis of the atheist's religion, which will take you nowhere.

It is a depressing soul-destroying religion, I plead with you to turn your back on evolution.

As I was writing this epilogue, I went downstairs for coffee and turned the TV on while taking a break. To my surprise, the home page for YouTube appeared and the first video was a clip titled, '*A 97-year-old Philosopher faces his own death*. A very unusual title as YouTube normally launches with my usual news channels.

I thought it strange for that title to appear. Curiosity won out, and I decided to watch it.

As this elderly philosopher with nearly a century of living on this small planet of ours was nearing the end of his days, his grandson recorded his views on life and his pending departure from this world, reasoning his words would be of great significance.

As a philosopher, he had written books expressing his views on subjects, one being death. He was an evolutionist and no doubt an atheist, as he didn't believe in God.

His philosophy on death, as outlined in his book, was that after we die, we no longer exist, it's all over forever, the end. The core belief of atheists.

Many Philosophers and academics mistakenly have this viewpoint.

However, despite this being a key belief for his entire life, when he was at the end of his mortal coil and the reality of his mortality began closing in, he started rethinking his stance.

Obviously when any academic writes about death, their views are based on their outlook at the time. If they are young and healthy, death often does not seem like a real prospect they are about to face, it's more of an academic debate.

In his book this philosopher questioned why people are afraid of death, as he reasoned in his mind that if there is nothing after death, you're not going to be afraid or unhappy…you're not going to exist…so there is nothing to be afraid of.

But after all these years, his once purely academic debate was now a reality, staring him in the face. His mortality was now facing him head-on, causing a litany of new questions to form in his mind, his thinking was changing.

This elderly man was now facing the realization that what was once a logical academic argument penned many years ago, was something he was no longer so sure of as the theoretical gave place to the reality that was rapidly closing in.

His views were changing. As he sat outside on his deck, overlooking the trees and nature, he mentioned how he was now seeing them in a new and different light. He explained it was like awakening to a different realm.

I would like to think perhaps he was seeing the Creator's hand in nature that he previously had been too busy in his academic world to appreciate, and was now finally observing the obvious with the blinders removed from his eyes. Nature was created, and there had to be a Creator.

Unfortunately, the video doesn't show the outcome of his reassessment of his coming demise.

What we do know is one day we will all face our own mortality. As a wise man once wrote, there are only two things you can be sure of, taxes and death.

This book proves the Big Bang and evolution are impossible.

There is a Creator, and regardless of what you may think, our existence doesn't finish when we leave this world.

Every time you eat chicken or eggs, take time to let them remind you of the most important discovery you will make in your life. The Chicken came first, he had to be created so there is a Creator. It's proven science.

Get to know your Creator while you can.

Colin B Noble can be contacted via email.

theyconnedyou@gmail.com

www.ingramcontent.com/pod-product-compliance
Lightning Source LLC
Chambersburg PA
CBHW031414290426
44110CB00011B/381